THIRD WORLD IN THE FIRST

One of the major cultural and economic issues facing both Australia and Canada concerns the governments' past and present failures to provide the 'first peoples' with appropriate development opportunities.

Elspeth Young contrasts the materialist development approach of both big companies and governments with the stress the Indian, Inuit and Aboriginal peoples place on husbanding natural resources.

Exploring why attempts to promote minority development have failed, whether models of sustainable development are applicable to remote area development as well as the crucial issue of self-determination, the book reveals the yawning gap between what people want and what governments are prepared to offer. The author argues that this gap can only be bridged by alternative approaches to development, centred on participation and the acknowledgement of these peoples' holistic sense of community.

A brief overview of the development impact on Botswana's 'first peoples', the Basarwa, extends the comparative approach to the issue of indigenous groups in general, be they in first or third worlds.

Elspeth Young is Reader and Director of the Graduate Program in Environmental Management and Development, National Centre for Development Studies at the Australian National University.

THIRD WORLD IN THE FIRST

Development and indigenous peoples

Elspeth Young

London and New York

First published 1995
by Routledge
11 New Fetter Lane, London EC4P 4EE

Simultaneously published in the USA and Canada
by Routledge
29 West 35th Street, New York, NY 10001

Typeset in Garamond by Solidus (Bristol) Limited
Printed and bound in Great Britain by
Biddles Ltd, Guildford and King's Lynn

British Library Cataloguing in Publication Data
A catalogue record for this book is available from the British Library

Library of Congress Cataloging in Publication Data
Young, E. A. (Elspeth A.)
Third World in the first: development and indigenous peoples/
Elspeth Young
p.   cm.
Includes bibliographical references and index.
1. Australian aborigines – Economic conditions. 2. Inuit –
Northwest Territories – Economic conditions. 3. Indians of North
America – Northwest Territories – Economic conditions. 4. Rural
development – Australia. 5. Rural development – Northwest
Territories. I. Title.
GN666.Y676   1995
307.1'412'0994–dc20   94-11463

ISBN 0–415–05543–1
0–415–11673–2 (pbk)

CONTENTS

PLATES

FIGURES

TABLES

PREFACE

What, people often ask me, could possibly be the connection between such far-flung parts of the globe as the central Australian desert, Canada's Northlands and Alaska's Arctic slope, and the Kalahari desert? Have I conducted research there purely because I, as a geographer, like travelling? And, once there, how can an Australian, born a native Scot brought up in Edinburgh's damp and blustery climate, possibly cope with interpreting some of the complex environmental and human interactions which provide them with their contemporary character? The answers to these questions are perhaps more straightforward than might be expected. Although travelling is certainly enjoyable and stimulating, the reasons for conducting comparative work in these places are academically important and enlightening. My interest in problems of remote development goes back to the beginning of my research career when, as an Edinburgh University undergraduate, I studied socio-economic change in a parish in Scotland's Northwest Highlands. Here the evidence both of the overpowering strength of new development – sheep farming, forestry and, more recently, the tourist industry – and the resultant dispossession of highland families and their transportation to Canada during the 'Clearances' clearly exposed the inequalities stemming from different perceptions of human values and aspirations. The industrial invasion of the clan lands by southerners was in practical terms the implementation of a colonial policy which has important parallels with elements of Australian and Canadian history.

Outback Canada and Australia share many common traits – their isolation from urban developed parts of the nation states to which they belong; harsh climates and rugged landscapes which pose challenges to human survival; their frontier images, drawing in resource developers, adventurers and those in search of alternative lifestyles from the outside; and most particularly their importance as the homelands of significant aboriginal groups, the 'first peoples' of modern day Australia and Canada. Along with such features goes a shared political history, derived from the former status of these countries as British colonies. In these circumstances it is hardly surprising that among the fascinating parallels which emerge from comparative research in these areas is that of modernisation and its impact on the aboriginal people.

The impact of development on Aboriginal people in remote Australian communities has formed the background to my research ever since 1978, when I first embarked on a research project on the Aboriginal economy. Studies of economic enterprises, such as retail stores, pastoral stations and arts and crafts have been an important emphasis of this work. Consultancies on Aboriginal land claims, coupled with studies of contemporary mobility and a growing but necessarily imperfect understanding of the personal and intergroup relationships which form the fabric of their societies have given me some insight into why such enterprises are very different in character from their counterparts in conventional non-Aboriginal society. Exposure of these differences in turn led to the central problems – the failure of non-Aboriginal people in general and government bureaucracies in particular to recognise the cross-cultural bases for Aboriginal responses to development; and their resultant failure to provide people in remote Australia with types of support which would allow them to achieve the lifestyles which they sought for themselves and their children. Not surprisingly these problems give rise to the two questions central to this study:

- Why have many of the attempts to promote aboriginal development failed?
- What alternative forms of development might be attempted?

Such seemingly simple questions become more complex as one's detailed knowledge is extended and it is here that the comparative method becomes important. Comparisons bring out broad trends and processes which otherwise are masked in the minutiae which inevitably preoccupy members of each individual community. Extending such comparisons beyond Australia to Canada has highlighted themes which otherwise might have remained disguised. In 1984 and again in 1989 I spent periods of six months in Canada. These visits allowed me not only to familiarise myself with the thinking of Canadian social scientists working on similar issues but also to spend some time in Northwest Territories and, in 1989, in Ottawa talking to government officials working in the aboriginal development policy area. The comparative dimension was enhanced in 1993 by a short sojourn at the University of Botswana followed by brief familiarisation visits to Basarwa communities in the Kalahari desert. The broad issues which emerged from this – basic questions of Basarwa identification, of assimilation in the cause of national unity, of the dominance of resource development at the expense of the welfare of the 'People of the Desert', and of the lack of a Basarwa political voice – persuaded me that a short overview of that situation would help to set the scene for what is essentially the crux of this text – aboriginal development in remote Canada and Australia, as the 'third world in the first'.

Essentially this book synthesises the results of my fieldwork research since 1978. Detailed descriptions of most of the results have previously been published. But broad theoretical comparisons have been less well developed, as have considerations of how the evidence fits into the wider scope of development studies. In keeping with much of the recent discussion in these types of inquiry the sustainable development model is employed as the key which explains basic differences in the development aspirations of aboriginal people and those responsible for the policies and programs which affect them.

With a time coverage of some sixteen years and the broad comparative nature of the research it is to be expected that the number of people whose ideas and stories have contributed to this book is enormous. In Australia they include the 'outback' Aboriginal people who taught me so much about their land and their lives – in Yuendumu, Mt Allan, Willowra, Ti Tree, Numbul-war, more recently in the east and north Kimberley, and more briefly in remote communities stretching from Bidyadanga in Western Australia, to Aurukun in Queensland, Fregon in South Australia and including the whole of the Northern Territory. Similarly Canadian aboriginal people in Fort Good Hope and Cambridge Bay not only opened my eyes to their way of life and how they interacted with the outside world but also made me feel at home. Indeed, only twenty-four hours in Cambridge Bay made me feel as if I had discovered an Arctic Yuendumu. My brief stay under the care of the people of D'Kar in Botswana also induced this feeling of homecoming.

In towns like Alice Springs, Darwin, Kununurra, Derby, Inuvik and Yellowknife members of aboriginal organisations and people working in government offices and in the private sector gave freely of their time, information and stimulating discussion. Canberra and Ottawa bureaucrats also, particularly in the Australian Department of Aboriginal Affairs/Aboriginal Development Commission/Aboriginal and Torres Strait Islander Commission and the Canadian Departments of Indian and Northern Development and Industry, Science and Technology were very generous both in interviews and allowing access to information. Many academic colleagues, with a breadth of experience which reflects the interdisciplinary nature of such work, have contributed through discussion, brain-storming and critical assessment. Among them Nugget Coombs, Fred Fisk, Fay Gale, Linda Ellanna and more recently Richie Howitt, Joss Davies and Helen Ross have provided important sounding-boards; in Canada the same roles have been performed by Peter Usher, Hank Lewis, Fran Abele and Frank Tough; and much more recently Sidsel Saugestad was my mentor in Botswana. Three former colleagues were of key importance – David Penny, for his early insights into Aboriginal Affairs and development in Australia; Charles Rowley, for his wisdom on so many aspects of the place of land acquisition and Aboriginal development; and Sally Weaver, another comparative Canadian–Australian fieldworker whose grasp of policy issues, wisdom and humour made my own efforts seem worthwhile. I deeply regret that because of their untimely deaths they are no longer here to see the final product.

Practical support for this work has also been crucially important. In Australia it has come from the Australian National University and the University of New South Wales, for whom I have worked since the late 1970s. These institutions supported my overseas study leaves in Canada in 1984 and 1989, and in Botswana in 1993. Visiting fellowships at the Australian National University in Human Geography and the North Australia Research Unit, at the University of Alberta's Boreal Institute and visitor status at the University of Botswana's National Institute of Research, Documentation and Development were vital elements in this work. Outside research funding in Australia has come from the Commonwealth Department of Aboriginal Affairs, Northern Territory Departments of Education and Community Development and, most recently, the Australia Research Council. I owe a great deal to the support of my colleagues in the Department of Geography aned Oceanography at the Australian Defence Force Academy who have made many useful comments during seminar presentations and who have been tolerant of the amount of time which has had to be devoted to completing this work. In particular I thank Anne Dunbar-Nobes and Daphne Nash for their meticulous work on creating annotated bibliographies on Canadian and Australian issues, Paul Ballard for his cartographic expertise and Julie Kesby for her cheerful assistance with computer crises, proof-reading and so many other detailed aspects of manuscript production. The final product, with all its errors and inconsistencies is my responsibility.

ABBREVIATIONS

ABTA	Aboriginal Benefits Trust Account
ACL	Arctic Cooperatives Ltd
ADAB	Australian Development Aid Bureau
ADC	Aboriginal Development Commission
ADM	Argyle Diamond Mine
AEDP	Aboriginal Employment Development Program
AFN	Assembly of First Nations
AGPS	Australian Government Publishing Service
AIAS	Australian Institute of Aboriginal Studies
ALC	Aboriginal Loans Commission
ALFC	Aboriginal Land Fund Commission
ALP	Australian Labor Party
ALPA	Arnhem Land Progress Association
ANPWS	Australian National Parks and Wildlife Service
ANU	Australian National University
AP	Anangu Pitjantjatjaruku
ASIG	Argyle Social Impact Group
ATSIC	Aboriginal and Torres Strait Islander Commission
AWS	Anangu Winkuku Stores
BFS	Business Funding Scheme
BHP	Broken Hill Proprietary
CACFL	Canadian Arctic Copperatives Ltd
CAEDS	Canadian Aboriginal Economic Development Strategy
CALM	Conservation and Land Management
CAP	Canadian Arctic Producers
CARC	Canadian Arctic Resources Committee
CCNT	Conservation Commission of the Northern Territory
CDEP	Community Development Employment Program
CEAC	Canadian Eskimo Arts Council
CEIC	Canada Employment and Immigration Commission
CEIS	Community Economic Initiatives Scheme
CLC	Central Land Council

COPE	Committee of Original Peoples Entitlement
CRA	Conzinc Riotinto Australia
CRES	Centre for Resource and Environmental Studies
CSIRO	Commonwealth Scientific and Industrial Research Organisation
DDA	Department of Aboriginal Affairs
DEET	Department of Employment, Education and Training
DEW	Distant Early Warning
DIAND	Department of Indian and Northern Development
DOGIT	Deed of Grant in Trust
DRIE	Department of Regional and Industrial Expansion
EKIAP	East Kimberley Impact Assessment Program
ESD	Ecologically Sustainable Development
FCNQ	Federation des Cooperatives du Nouveau-Quebec
GDP	Gross Domestic Product
GNP	Good Neighbour Policy
GWNT	Government of Northwest Territories
HTA	Hunters and Trappers Association
IDC	Inuvialuit Development Corporation
IGC	Independent Grocers Cooperative
ISTC	Industry, Science and Technology Canada
ITC	Inuit Tapirisat Canada
IUCN	International Union for the Conservation of Nature and Natural Resources
KDT	Kuru Development Trust
KLC	Kimberley Land Council
NARU	North Australia Research Centre
NCC	Native Capital Corporation
NEDP	Native Economic Development Program
NGO	Non-Government Organisation
NIB	National Indian Brotherhood
NLC	Northern Land Council
NSW	New South Wales
NT	Northern Territory
NTOC	Northern Territory Open College
NWT	Northwest Territories
NWTCBDF	Northwest Territories Cooperative Business Development Scheme
RAD	Remote Area Dweller
RADP	Remote Area Development Program
RCIADIC	Royal Commission into Aboriginal Deaths in Custody
Special ARDA	Special Agricultural and Rural Development Agreements
TAFE	Technical and Further Education

TFN	Tungavik Federation of Nunavut
TGLP	Tribal Grazing Land Policy
UBC	University of British Columbia
WA	Western Australia
WCED	World Commission on Environment and Development
YBE	Yirrkala Business Enterprises

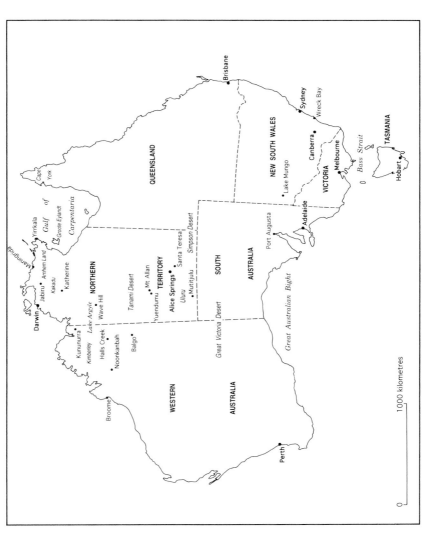

Frontispiece 1   Australia: main towns and places referred to in text

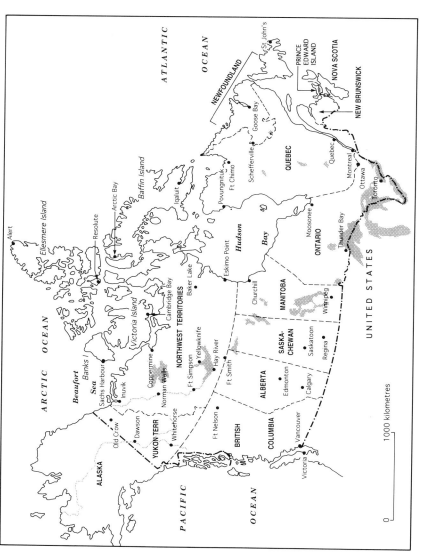

Frontispiece 2 Canada: main towns and places referred to in text

1

DEVELOPMENT AND ABORIGINAL PEOPLE IN REMOTE CANADA AND AUSTRALIA
An overview of the main issues

INTRODUCTION

The ideology of development has, for decades, formed the foundation of much of our thinking about the present and future state of humankind. As a process whereby the human condition is improved through more effective use of resources – environmental, social, economic and political – it has generally been interpreted as advantageous. All peoples, it is assumed, should strive for development and in the process should be able to reap lasting benefits for themselves and for their children. Growing evidence of resource depletion and misuse in the development cause – land degradation, the pollution of waterways, the overuse of non-renewable resources – and of the increasing width of the gap between rich and poor nations and wealthy and poverty-stricken individuals has led many people to question this assumption. Nevertheless most people in the industrialised and industrialising world still perceive that the advantages gained through increasing material wealth far outweigh such disadvantages.

People from other worlds, those of the 'third world' where the stranglehold of industrialisation is not absolute and of the 'fourth world', the indigenous minorities[1] who have survived the industrial onslaught on their homelands, may well perceive development rather differently. They clearly recognise that, in its conventional form, the development process can bring benefits, such as redressing socio-economic disadvantage, providing better access to opportunities arising through technological advancement and enhancing both political and economic power. But they also recognise its negative implications. These include not only environmental degradation but also cultural and social destruction and the rupturing of the intimate relationships between human beings and their natural environment. Since for them development is a process which promotes not only their economic advancement but also their social and cultural vitality and which emphasises long-term sustainability rather than short-term gain such costs may be

1

unacceptable. This view of development will often conflict with the industrial ideology.

The essential differences between these two concepts of development can be highlighted in a variety of situations. These include comparisons between industrialised countries and the developing countries which provide them with their resource base for growth; and, within developing countries, comparisons between urban and rural sectors. The development priorities of politically powerful groups compared to those of minorities also reveal these differences. Arguably they are nowhere more obvious than within nation states where, despite decades of pressure exerted through industrialisation, indigenous minority groups – the 'first' or 'aboriginal' peoples – cling tenaciously to their land and their culture. This situation, examined primarily in the context of the nation states of Australia and Canada, provides the focus of this book.

Attributes shared by Australia and Canada include not only their historical and political inheritance as former outposts of the British Empire and contemporary members of the Commonwealth but also their vast size, the inhospitable nature of large tracts of their environment, their wealth in non-renewable resources and the small size and high degree of concentration of their population. In both countries this has, among other things, resulted in an attitude to development which stresses resource exploitation and can result in human and environmental detriment. Nowhere is this more apparent than in their remoter regions, the Canadian northlands and the tropical wetlands and deserts of Australia's north and centre. And in both countries it is in those very regions that the presence of the 'first peoples', the Inuit and Indians of Canada and the Aboriginal and Torres Strait Islander peoples of Australia, is most marked. Differences between their perceptions of development and those of the more recent Australian and Canadian settlers from 'south' are inevitably highlighted. The title of Berger's (1977) report on the Mackenzie Valley Pipeline Inquiry – *Northern Frontier, Northern Homeland* – expresses the essential character of these differences. To the incomers the Canadian north is a frontier land, with resources to be exploited for national material benefit and by whatever means might prove to be commercially most profitable; to the northern Indians and Inuit it is quite simply their homeland, the place to which they remain inextricably linked in social and cultural terms and for which they are the curators for future generations. Similar comparisons have been made in Australia. Aboriginal relocation of population back to remote parts of the desert or Arnhem Land has been expressed by government officials as an 'outstation' movement, but by the Aborigines themselves as a return to 'homeland'; the former group, predominantly urban dwellers in towns such as Darwin or Alice Springs, see these places as part of the periphery but to the latter they are the core. Coombs *et al.* (1989), in their discussion of development challenges in the East Kimberley region of Western Australia also contrast the conventional philosophy through which development

has generated material goods of a diversity and on a scale unprecedented in human history, ... has ... provided the basis for a life both
rich and satisfying [and] continues to underlie the economic policies of
most governments

with an Aboriginal philosophy of development which is governed by

a set of beliefs which carries the authority of religion and ancestral law
[which emphasises] the continuity of their present experience with that
of the past and seek[s] meaning for the present in terms of the past.

The expression of such parallel themes in relation to physically contrasting
areas such as the Canadian north and Australia's central deserts is not
accidental. It stems partly from increasing questioning by members of the
industrialised society of the conventional approach to development, a
questioning which has resulted in the current emergence of the need for
policies promoting ecologically sustainable development. It also reflects the
growth in political and economic power achieved by aboriginal peoples over
the last two and a half decades, a growth which has been based on recognition
of their plight (Burger, 1987) and of their rights to self-determination
including the granting of land rights. The latter change, in particular, has been
highly significant in remote areas of both Canada and Australia. These
themes have recently been brought together by writers such as Knudtson and
Suzuki (1992), who have illustrated how, through the *Wisdom of the Elders*
(the first peoples), resource use could have been sustained much more
effectively than has occurred under the custodianship of industrialists. The
designation of 1993 as the International Year of Indigenous People is further
proof of a more enlightened and sympathetic understanding of these issues.

At the same time these groups of people remain by any form of measure
the most disadvantaged within their nation states. Low incomes, high levels
of unemployment, lack of formal educational qualifications, poor health and
substandard housing conditions all confirm this situation (DIAND, 1980;
Wonders, 1987; Young, 1988c). Signs of a more sympathetic understanding
of the cultural and environmental knowledge of aboriginal peoples are not
enough to ensure their future survival. It is also necessary to find a means for
their development, in ways appropriate to their lifestyles and aspirations.
This study, through examining aspects of development affecting aboriginal
peoples in remote parts of Canada and Australia, highlights common issues
which must be taken into account in the formulation of policies for their
future support. Specific questions to be addressed in both regions include the
following:

- Why have many of the attempts to promote aboriginal development
  failed?
- What alternative forms of development might be attempted?

3

DEVELOPMENT: THE BROADER CONTEXT

The ideology of development has, as Coombs *et al.* (1989: 5) note, been 'a powerful intellectual and political force which … has changed and is changing the face of the world'. It is therefore important that, as they further comment, the common assumption that development is 'good for everyone' is properly questioned and assessed. This is particularly necessary when this ideology is being applied in a cross-cultural context, as in the case of indigenous minority groups in the industrialised world. Current aboriginal development policies in both Canada and Australia are products of a historical process which, in turn, reflects the changing emphases of development ideology through time. Processes such as decolonisation, the internationalisation of capital and labour and the dominance of global trade systems have all influenced aboriginal development and have produced outcomes which strongly parallel those observed by social scientists examining 'third world' development. A detailed analysis of development thinking, such as those conducted by Blomstrom and Hettne (1984) or Forbes (1984), lies beyond the scope of this study. However, within recent decades, changes in development thinking do highlight some important points of relevance to the situation of aboriginal peoples.

Development has commonly been described as a process leading to modernisation, whereby societies disadvantaged in terms of living standards and material wealth reach socio-economic levels perceived to be acceptable to society as a whole. As this implies, modernisation is a relative term and one which, like development, is often too loosely used (Mabogunge, 1989: 35). Conventionally, however, it is interpreted primarily within an economic context, measured by advances such as increase in income, participation in wage labour and growth in material wealth. This view of development and modernisation reflects the perceptions of those belonging to rich sections of the industrial world, people who themselves have adopted such an approach in their pursuit of a better life. This has led them to label other societies whose members exhibit different priorities as 'primitive', 'backward' and 'archaic'. Development theories of the 1950s and 1960s, incorporating concepts such as progression through successive stages to reach 'take-off' (Rostow, 1960), or the diffusion of the benefits of modernisation through 'trickle-down' (Hirschman, 1958) or movement from the 'core' to the 'periphery' (Myrdal, 1963; Friedmann, 1966) all reflect such perceptions. These theories can be described as sharing a 'top-down' approach to development, an approach which stresses the overwhelming importance of industrialisation, monetisation and the adoption of a belief in the need for resource exploitation on a large scale. Such an approach assumes that those not yet subscribing to such beliefs would be indoctrinated until their views were assimilated with those of the industrial society. Assimilationist approaches of this type are a vital part of the foundations of aboriginal policy in both Canada and Australia and, many would argue, still prevail.

During the 1960s and 1970s this conventional view of development, with its mainstream emphasis on economic elements was, as Forbes (1984) discusses, challenged both by those directly affected by the process (members of 'third world' nations) and by academics and practitioners in the field. Questions were posed for a number of important reasons. These are as follows.

Firstly, transformation induced through economic processes frequently did not lead to development within all sectors of society. Rather than the advantages of industrial development – the jobs, higher incomes and educational opportunities – 'trickling down' from growth centres to people in the periphery, the two components were increasingly separated; more highly skilled/innovative individuals flocked to the growth centres, participating in rural urban migration often on a scale sufficiently massive as to cause huge social disruption; and in the process, polarisation of society into rich and poor, high and low status classes was ever more marked. The concept of 'marginalisation' – the consignment of peripheral societies to perpetual poverty and disadvantage in terms of means of sustenance and provision of services – describes such a process. The cross-cultural aspects of that situation, as Park's original (1937) definition of 'marginal man' as 'one whom fate has condemned to live in two societies and in two, not merely different but antagonistic cultures' implies, make the marginalisation approach particularly apposite to analysis of aboriginal development. In Australia it has, for example, been fundamental to Drakakis-Smith's (1980) discussion of the situation of Aboriginal town-campers in Alice Springs and has been more recently applied to people living in remote rural locations (Young, 1991b).

Secondly, economic development cannot stand alone but is firmly embedded within social and political contexts and is a product of its history. Cultural attributes and behavioural norms influence how people perceive the changes which they are being encouraged to adopt. Industrialisation, with its emphasis on regular work, regimentation and, at least at managerial levels, striving to beat one's fellow humans in the game of life, may well conflict with modes of behaviour which stress flexibility, choice, sharing and reciprocity and place a high value on community rather than personal advantage. Failure to acknowledge such a conflict may lead to enormous expenditure of human and economic resources on projects of little long-term benefit to those in need. Porter *et al.*'s meticulous analysis of an Australian aid project in Kenya (1991) presents one such example. There are many more. Solutions lie not merely in recognising the importance of social and cultural elements in development, but also in finding alternative approaches which incorporate them. The broadening of the development debate in the 1970s with contributions from social scientists other than economists (for example, sociologists, geographers, political scientists) reflects this changing attitude.

Development economists also began to express these views. Thus Fisk (1982), in discussing development and aid to Pacific countries states clearly 'economic aspects are interdependent with the social, political and strategic aspects, and require to be seen in that context if sense is to be made of them'. More recently Ross and Usher (1986), in their stimulating discussion of formal and informal economies, have highlighted the importance of viewing economic development in this wider sense. As they suggest, such a holistic approach is appropriate for all societies, whose economies lie partly in formal spheres such as big business or government and also in informal spheres such as household structures and community organisations. It is particularly important for societies such as those to which aboriginal peoples in Canada and Australia belong, where economic activity is generally small scale and where family relationships continue to play an important role. Government policy implementation in both of these countries still fails to take sufficient notice of these characteristics.

Thirdly, 'top-down', economically oriented large-scale development processes generally fail to consult effectively with those undergoing development to gauge what their aspirations for the future might be. Even more rarely do they then enter into negotiations about the implementation of development and the sharing of its benefits. This lack of participation will often, as Goulet (1980) has commented, lead to obvious conflicts between these views; development may, as a result, have a negative impact. Thus in subsistence farming communities such as those in Melanesia, the introduction of cash crops may be encouraged as a vital step towards monetisation and industrialisation; however such development may undermine basic systems of human survival, disrupting land tenure, increasing erosion and nutrient depletion in the soils, and decreasing the flexibility in economic strategies which enabled people to maintain the essential balance with their environment (Lea and Curry, 1988). Similarly, incorporation of people into industrial wage labour can destroy their subsistence base, increasing their dependence on cash and leaving them more vulnerable to uncontrollable vagaries of the market and external economic forces. As analyses of the impact of the Bougainville copper mine in Papua New Guinea and of the social, political and economic traumas which have recently occurred there show, proper and responsible consultation in the initial stages might well have prevented the present tragic confrontations (Connell, 1991).

Goulet's definition of participation includes the 'organised efforts to increase control over resources and regulative institutions in given social situations' and can occur through three possible channels – induced by an authority above, generated from the 'grassroots' below, or stimulated through the encouragement of an interested outside agent. Alternative ways of achieving a more satisfying lifestyle can often emerge through this type of communication. Unfortunately, as Porter et al. (1991) comment, government authorities rarely manage to achieve this type of dialogue, even when they are aware of its value. Aboriginal development, overwhelmingly influenced by government agencies, is clearly disadvantaged by this omission. However, as some recent studies in Australia (for example, Coombs *et al.*, 1989; Crough *et al.*, 1989; Crough and Christopherson, 1993b) and Canada (for example, Wolfe, 1988; 1989) have shown, effective participation can form the foundation for academically oriented research projects and the eventual outcomes are certainly likely to be more valuable than they might be otherwise.

Finally the political and economic aspirations of nations in the capitalistic world influence development elsewhere. In general their governments and private companies view development from the perspective of their own advantage, either directly through gaining control over resources, or through maintaining a political balance which enables that control to be exercised. Thus partial industrialisation, sometimes referred to as 'un-even development' (Forbes and Rimmer, 1984), may well reflect deliberate strategies imposed through the operation of external forces, on either a national or a global scale. Preventing the growth of secondary processing after resource extraction, hindering the allocation of wage levels which provide people with adequate living standards, refusing to provide family housing so that the workforce consists largely of single men or women with more limited service needs, are all common practices in this situation. And on a global level the artificial maintenance of trade agreements and barriers also deliberately slows down the rate of change.

Such deliberate strategies led in the late 1960s to the concept of 'underdevelopment' and 'dependency', clearly outlined by Frank (1967). Using Latin America as his example, he discussed how that region had remained internationally peripheral, an economic satellite of the United States; and also how regions within Latin America, notably inland areas with poor communications and sparse populations, had been maintained in an underdeveloped state by power and control exerted by large and wealthy organisations operating from core cities such as Rio de Janeiro and Santos. The theory of underdevelopment had attractions at the time, principally because it highlighted the significant effects of global and national political and economic frameworks, in creating what Buchanan (1977) called not separate processes of development and underdevelopment but a single process made up of both these elements. It did however draw severe criticism, summarised in Forbes (1984: Ch. 4). As he showed, dependency theory did not explain why some countries at the periphery were underdeveloped through exploitation of their resources for the benefit of a colonial power (for example Indonesia) while a similar process resulted in development in countries such as Australia. Forbes also points out that dependency theory failed to recognise the importance of social and cultural influences within societies affected by underdevelopment and his arguments about the vital effects of class systems and the strength of resistance to colonialism are cogent and persuasive. Aspects of dependency theory are nevertheless useful in considering the frameworks within which aboriginal development in the nation state operate.

These questions have helped to extend the view of development well beyond that initially engendered by the economic-based theories of earlier times. Goulet's (1973) broad description of development as a process which offers 'the opportunity to live full human lives' expresses this more holistic and realistic definition. Within this these questions imply: firstly, that successful development must take into account the aspirations of those 'being developed' (the clients), as well as the desires of the 'developers' (the providers); secondly, that development occurs within a human society whose cultural and social attributes are vital influences on the process; and thirdly, because of the global linkages between the developed and developing world, that elements such as resource base and population growth have to be considered on an international as well as on a national or regional scale if development is to occur.

None of these ideas and definitions provided practical answers to what many people would see as the prime challenge of development – how to find some way of breaking the nexus of power exerted by the rich developed world/groups over poorer nations and sectors within society. Such a break, which would imply the exertion of much greater levels of self-reliance and self-determination on the part of those experiencing development, is obviously hard to achieve. However a number of other views of development have attempted to take up some of the ideas expressed above, notably in relation to participation, consultation and negotiation, in subscribing to the holistic socio-economic nature of the process and in the need to give priority to people on the periphery, often the rural-dwellers. Needs-based planning, often called 'grassroots development' is one example of such an approach. In reflecting the feelings and wishes of those whom development is intended to help it should provide a more appropriate framework for attaining their goals. This approach was fundamental to the work of geographers involved in planning for provincial development in Papua New Guinea in the 1970s and early 1980s when that newly independent country was still strongly committed to implementing a national rural development policy (for example, Howlett *et al.*, 1975; Carrad *et al.*, 1982; Simpson, 1987).

By its very nature needs-based planning is often perceived to be synonymous with rural development planning. While theoretically this need not be so, the reality is that peripheral regions aiming for development will often be rural. Hence the approach to rural development, discussed at length by Chambers (1983), is essentially a needs-based approach which 'puts the last first'. It may also be a community-based approach whereby development must not only take account of community needs and aspirations but must also consider the community from a holistic (social, cultural, economic and political) viewpoint. Such approaches to development apparently have much to offer. People affected by the process know that they have had a direct input into the planning of projects, and a better opportunity of participating in their implementation. They are also better able to tailor planning needs to their own society, defined not as a loose conglomeration of a number of separate and barely connected elements but as a complex and intricate web of inter-related components.

Despite these advantages needs-based, rural and/or community development planning has proved difficult to implement at either national or regional scales. Crittenden and Lea (1989) comment that not only is it difficult to get bureaucrats and politicians to commit themselves to this approach but also that the practitioners who are charged with facilitating it themselves influence the outcomes through both overt and hidden agendas. Thus the needs expressed may not really be those stressed by the people. In Papua New Guinea, for example, the emphasis appears to have changed in favour of greater centralisation of development at the core, and the perception that national rather than regional advantage should receive principal attention. Internal pressures from urban-based elite groups and external stresses resulting from the need for economic viability in a global sense have been too strong. Thus the desires of the village community may well be passed over in favour of large-scale resource development, such as the Bougainville copper mine or the Porgera gold mine. Such a development policy can, as Papua New Guinea has shown, be disastrous at both village and national government levels and can even have widespread destabilising effects. A main dilemma in development is the successful merging of these small- and larger-scale goals in order to provide a 'full human life' which satisfies everyone. Goulet (1980) at least considers that this should be possible, although he acknowledges its complexity. It is certainly a development problem confronting aboriginal peoples in remote parts of Canada and Australia, faced with external pressures to permit mining on their land even when they know that the local benefits may be quite small and the local costs quite large.

As this brief overview indicates our interpretation of the meaning of development has, over the last half century, become increasingly complex. Given the changes which have been occurring within global political and economic systems and our increasing firsthand knowledge of other cultures and nation states, this is perhaps not surprising. However additional factors have now become a vital part of the development debate, factors which stem from worldwide concern over the rate of resource depletion, the global warming effect caused by greenhouse gas emissions and the continued high rates of human population growth in many societies. These problems have led to a welding of environmental issues, including both the notion of conservation and that of careful husbandry, to the other facets of development – its economic, social and political components. It has also led to increasing acknowledgement that successful development can rarely occur in the short term, but will inevitably take a long time to implement. During that process it may well be that both the method of implementation and the ultimate goals of development will change. These ideas have been incorporated into the ideology of sustainable development, which in the early 1980s argued for a global agenda for change. The World Commission on Environment and Development (WCED) has defined sustainable development as a process which 'meet[s] the needs of the present without compromising the ability of future generations to meet their own needs' (WCED, 1987: 43). This, as has commonly been stressed, is a tall order. Indeed, it has increasingly been suggested that the two terms 'sustainable' and 'development' are completely incompatible, and that, as global recessions and unemployment bite even more deeply into the fabric of industrialised society, the concept of sustainability, however much we may believe in it in the long term, should be discarded.

WCED was commissioned by the United Nations to examine, amongst other things, how:

*long-term environmental strategies for achieving sustainable development could be most quickly set in place;
*there might be greater co-operation between countries at different stages of economic and social development so that they could work out common objectives supporting people, resources and environment;
(WCED, 1987: ix)

Such aims are of universal relevance and many industrialised countries, including both Australia and Canada, have now spent considerable time and effort on discussions on what sustainable development means, and how it might be achieved. Australia's national ecologically sustainable development (ESD) policy is the product of over two years of intensive discussion between government, scientists, academics and others within a series of working parties set up to examine prime components of development such as agriculture, mining, tourism and transport and communications (ESD, 1992). Over four hundred recommendations from these reports (ESD, 1991) have been presented. However while many in the Australian community acknowledge that there is a need for policies based on ESD, the adoption of these recommendations still awaits government commitment. Arguments over the economic losses which will be experienced in the cause of ecological sustainability continue to rage. The Canadian government has not as yet gone through a similar process on the national scale, although discussions, particularly in relation to northern regions, have been both intensive and productive for a number of years.

Sustainable development is no less relevant to 'third world' countries. Indeed, with their high rates of population growth, some would argue that it is all the more important. Barbier (1987: 102) states that:

'real' improvement cannot occur in Third World countries unless the strategies which are being formulated and implemented are environmentally sustainable over the long-term, are consistent with social values and institutions, and encourage 'grassroots' participation in the development process.

Such sentiments are, theoretically at least, beyond reproach. However the reality may be somewhat different. Third world development is generally resource based – mining, forestry, intensive tropical agriculture, or promoting attractions to gain the tourist dollar – and hence resources come under ever-increasing pressure. As the people's expectations for higher living standards increase, their own demand for resources will also rise. And philanthropic statements from the industrialised world about cutting down on resource use for the global good may cut little ice when people perceive the yawning size of the gap which separates them from those who are already rich.

The situation of aboriginal peoples has not gone unnoticed in the sustainable development debate. The World Commission itself commented on their plight and their potential contribution, making a special plea for their preservation as a vital component of the social and cultural diversity of human beings, and noting that their intimate knowledge of the environment and their tradition of resource husbandry make them uniquely qualified to practise sustainable development, and teach others some of their techniques (WCED, 1987: 114–16). In the Australian context Zarsky (1990) also specifically comments on the importance of recognising Aboriginal inputs and interests in ensuring the sustainability of land and resources. Canadians also, have highlighted the aboriginal peoples' contribution to sustainable resource use in their northern lands. Usher (1987a) has presented cogent arguments supporting aboriginal wildlife harvesting as a sustainable economic activity. Duerden (1992), while accepting that when such harvesting is done primarily for sustenance rather than for commercial gain this seems to be possible, also warns that some other forms of aboriginal resource use may not be sustainable. They and others acknowledge that the concept of sustainable development should be important for all northerners, both aboriginal and non-aboriginal, because it equips them with the means to counteract the problems stemming from the 'boom and bust' approaches of the past (Pretes and Robinson, 1989a).

Comments on the inherent sustainability of aboriginal resource use, coupled with recent popularisation of knowledge about aboriginal peoples' relationships with land, can create the impression that they did not change the environment. Such an assumption in either Australia or Canada would be quite erroneous. In Australia the disappearance of giant marsupials, and, above all, the use of fire to transform the landscape into something of greater utility for human sustenance bear witness to this. However in both Canada and Australia such changes occurred within human control, and were implemented within a society which, because of its knowledge of the

11

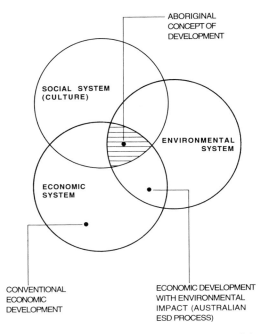

Figure 1.1 The sustainable development model
Source: Adapted from Barbier, 1987

environment, maintained that control. It is this that makes aboriginal resource use sustainable when compared to the forms of use introduced by non-aboriginal settlers.

How, then, can sustainable development be implemented? Here it is necessary to consider what an appropriate model of sustainable development might look like. According to Barbier (1987) development consists of three main systems: the biological and resource system; the economic system; and the social system. Sustainable development occurs where these three systems overlap; it thus contrasts markedly with conventional economic development which occurs solely within the economic system (Figure 1.1). The key to sustainable development does not lie in each system on its own, but rather in the interaction between all three. This holistic viewpoint accords well with dominant characteristics of aboriginal society, in which the people (the social system), their means to survival (the economic system) and the environment (the biological and resource system) are integrated so closely that none of these elements can usefully be studied in isolation. This model provides a valuable framework for understanding not only what the characteristics of aboriginal development might be, but also why so many attempts to promote development have failed. In simple terms, what aboriginal people have needed, and still need, are support systems aimed at total development, the characteristics of which are the same as those described for sustainable development; what they have commonly been offered have been policies and programs appropriate only to conventional economic or social development.

THE ABORIGINAL DEVELOPMENT CONTEXT: AN ILLUSTRATION FROM SOUTHERN AFRICA

As the preceding brief overview suggests understanding how development affects aboriginal peoples is a complex business. It involves consideration of environmental, economic, social and political factors, operating not in isolation from each other but integrated to form a complicated holistic process. In its contemporary form it is also the product of historical processes which still affect its characteristics to a significant extent. When the aboriginal groups concerned are part of an industrialised nation state, as in Australia or Canada, these complexities become particularly marked and any analysis aimed at providing detailed reasons for development's failures and successes involves some deconstruction of the whole into its separate parts. The resultant fragmentation detracts from the totality of the real life situation. The following case-study, a summary of some of the development experiences of contemporary Basarwa, or Bushmen peoples in Botswana with comparisons from neighbouring parts of Namibia (Figure 1.2), presents an integrated picture which highlights many of the common themes arising in subsequent analysis of aboriginal development in Canada and Australia.

The setting

Basarwa are distinct from other residents of Botswana in a number of ways. Most are remote dwellers, living within or on the edges of the Kalahari desert; they, the N/oakhwe or 'red' people are ethnically distinct from other 'black' or 'white' Batswana; they have retained into recent times at least the essential social and economic characteristics of a hunter-gatherer society; their languages belong to the Khoisan linguistic group, distinguished by phonetically complex click sounds; and they are the first peoples of that country, occupiers of the desert for many generations preceding the incursions of black African herdsmen or European farmers. Today they exhibit all the classic signs of a group bypassed in the development process – lack of political power, low socio-economic status, poor health, limited education and exhibiting signs of social strain such as alcoholism and high rates of imprisonment. Why?

Before becoming independent in 1966 Botswana was the British protec-torate of Bechuanaland, a land-locked arid region of southern Africa which was sparsely populated and apparently lacked valuable natural resources. Its affairs were dominated not only by Britain but also by its neighbours,

13

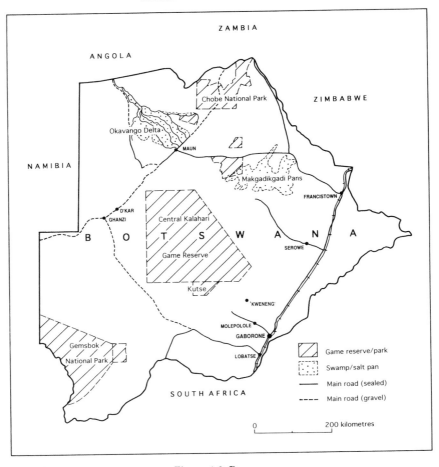

Figure 1.2 Botswana

particularly South Africa from which it was administered through the regional centre of Mafeking. Cattle, raised both for commercial and subsistence purposes, provided its economic focus and most of its small population, then numbering less than 600,000, belonged to Bantu-speaking peoples who had practised agriculture within Botswana for over a thousand years. Although these groups, including the Batswana and Bakgakagadi, must have frequently intermarried with the original Basarwa inhabitants their advanced technology allowed them to dominate the 'first peoples', take over control of their lands and exploit their labour. At independence the Basarwa were therefore at the bottom of the heap, socio-economically disadvantaged, largely landless and treated as slaves.

14

In 1966 the newly independent country of Botswana declared itself to be a non-racial state, where any practices amounting to apartheid and separate development would be abhorred. This highly democratic stance, which confers equal rights on all Batswana, clearly set the country apart from its neighbour South Africa. Coupled with its political stability and the richness of its recently discovered diamond deposits, it has helped Botswana to progress from being one of the poorest African nations to being one of the richest in the sub-Saharan region. In economic terms it is now firmly ensconced as a middle-income developing nation, an important exporter of diamonds and beef and with an increasingly important tourist industry. This is an admirable achievement. However, this rapid development has had its costs. Firstly, in emphasising that all people in Botswana have an equal right to share in the opportunities available it has discounted the particular needs of those who were already disadvantaged. It also assumes that everyone will want to take up these opportunities. It therefore follows an assimilationist line. Secondly, polarisation and associated marginalisation appears to be occurring. The distribution of this new found wealth is highly skewed. As Good (1993) has highlighted, Botswana has a remarkably wide gap between the highest and lowest 20 per cent of income earners (a ratio of 23.6:1, compared to 6.9:1 in Ghana according to United Nations figures in the mid 1980s). Finally Botswana's dependence on resource exports shows little sign of decreasing and external market forces thus exert strong control over her economic development. How are people in Botswana's remoter regions, in particular the Basarwa as a rural-dwelling minority group, faring under such circumstances?

Identification and Basarwa socio-economic status

As an ethnic group Basarwa are disparaged by most other Batswana. This attitude is well entrenched. It is based on the original perception that Basarwa hunter-gatherers, unlike pastoralists and farmers, did not 'own' the land and could therefore be displaced from their territory with impunity. Following that displacement many Basarwa were forced to work for the incoming cattle herders for little or no reward and were essentially treated as slaves to be disposed of at the pastoralist's will. Their cultural separation from other Batswana excluded them from conventional means of access to resources. According to the strictly hierarchical nature of Batswana society, firmly established during the nineteenth and twentieth centuries, control over vital resources such as land, water and property rights were determined by relationship to ruling families, the *kgosi* (chief) and his relatives (Data and Murray, 1989; Saugestad, 1993a). Such relationships were recognised for other minority groups such as the Bakgalakgadi but not for the Basarwa. In a country where wealth and status is still to a large extent measured by cattle ownership this means that the Basarwa are judged to be inferior. The persistence of such attitudes raises important human rights questions (Mogwe, 1992), and accords with the conclusion that failures in the development of hunter-gatherer peoples can be attributed more to socio-political factors than to economic causes (Bieslele *et al.*, 1989).

According to the United Nations Working Group on Indigenous Populations and the International Labour Organisation an indigenous minority group consists of those people who are descended from the first inhabitants, who are in a minority and do not control the national government, and who differ culturally from those who came later (Saugestad, 1993b). These criteria clearly fit the Basarwa. However, because of Botswana's non-racial political stance which prevents the official recognition of the ethnic identity of any separate population group, the Basarwa's status as the indigenous minority, the 'first people' of the country, has not yet been acknowledged. In 1993, the United Nations Year of Indigenous Peoples, Botswana's Minister for Lands, the department responsible for remote area development and hence most concerned with the Basarwa, justified his country's lack of involvement in the celebrations by stating that, in his and his government's view, all Batswana were indigenous (Saugestad, 1993b). Such sentiments, as others such as Wily (1978), Hitchcock and Holm (1991) and Good (1993) have commented, while understandable in the light of Botswana's determined and admirable stand against apartheid, must undermine efforts to raise Basarwa socio-economic and political status.

Such policies have some important practical consequences. Botswana's national census does not officially identify any minority groups through conventional means such as ethnic identity questions. There are therefore no statistics which can indicate differing levels of socio-economic status on an ethnic basis. Basarwa demographic and socio-economic characteristics therefore have to be assessed by less direct methods. Their current population is estimated at around 50,000 (about 3 per cent of the national population of around 1.3 million). Hitchcock's (1992a) figures, adapted by Saugestad (1993a), show them to be predominantly rural dwellers, with about 75 per cent of their total population being classified as Remote Area Dwellers (RADs), people who live in small settlements outside the conventional village-based local government structure. Since, as Good (1992 and 1993) has suggested, rural development in Botswana has been neglected and the already large income disparities between urban and rural workers are increasing this implies that Basarwa must be socio-economically disadvantaged.

Vast differences in the level of cattle ownership have also led to marked income disparities within the rural population. Here those who do not own cattle, a group to which many Basarwa belong, are the poorest, often with cash incomes derived entirely from drought relief, only 1176 pula (about $A600) per year in 1993. Basarwa labourers working as herders at cattleposts belonging to others have until recently been paid mainly in kind, with some access to milk and meat and provision of rations or clothing rather than in cash wages. This exploitation of Basarwa labour has been accompanied by the creation of a dependent relationship. As Guenther (1979), for example, has recorded for the Ghanzi farms region in the western Kalahari, the enforced displacement of Basarwa by the encroachment of pastoralism into much of their traditional country made them dependent on the cattle industry. The rewards which they received for their work sustained their families and also enabled them to continue living on or close to their traditional lands.

Basarwa, also, because of their geographical isolation and their social and political subjugation, have lacked educational opportunities, particularly at post-primary levels. Like their fellow rural dwellers those who have attended secondary and tertiary institutions have had to leave home and board in hostels in social environments which they have found hard to tolerate. Such practices, along with the institution of a core curriculum which is not sensitive to cultural differences, has helped in the rapid breakdown of Basarwa culture and identity (Mogwe, 1992). Their displacement from the land to centralised sedentary communities on cattle stations, government settlements and on the fringes of towns such as Ghanzi has also affected their health status. Resultant dietary changes and high alcohol consumption have been major factors. These deficiencies in education and health must undermine efforts to improve socio-economic status.

Gender differences in socio-economic status within the Basarwa community are also important. Evidence presented by Loermans (1992) indicates that while Basarwa women appear to have occupied a relatively egalitarian position in traditional society, the incorporation of Basarwa men into pastoralism as herders has increasingly marginalised them in contemporary society. Basarwa men have better access to wage jobs and own most of the property, including stock. The women thus have virtually no chance of having their claims to land and water rights recognised, and, as Hunter *et al.* (1990) show, their unique traditional knowledge and skills have been ignored in fields such as wildlife management and conservation where their expected contribution could be considerable. Loermans's succinct description of the difficulties encountered by a group of Basarwa women trying to establish their own enterprise based on the cultivation of *veld* foods highlights the problems faced by women who have to approach a bureaucracy dominated by masculine attitudes to development.

Explaining Basarwa socio-economic disadvantage

Reasons for Basarwa socio-economic disadvantage, implied above, include: their concentration in remote parts of the country; their cultural distinctions which do not accord well with assimilationist policies; their lack of access to land and resources; and conflict between the interests of Basarwa people and those which further Botswana's national development aims.

17

Remoteness

Remote parts of Botswana suffer in part because of simple distance factors – northern and western areas of the country have few all-weather roads, poor telecommunications and are very sparsely populated. Although since independence significant improvements have been made through government programs such as the Remote Area Development Program (RADP) (see below), service delivery, whether in education, in health, in agricultural extension or in financial and banking services remains much poorer than in southern and eastern parts of the country. Rural Batswana are all disadvantaged by these circumstances and young people in particular increasingly respond by leaving rural areas and moving to eastern towns such as Francistown and Gaborone in search of better conditions. Arguably Basarwa are doubly disadvantaged because of their particular attachment to the most remote areas such as the lands now included within the Central Kalahari Game Reserve (CKGR). They interpret the word *'tengnyanateng'* (Setswana, 'remote') in the opposite way from that commonly used in urban society:

> If by *'tengnyanateng'* it is meant that we are far away from Gaborone; Gaborone is also far away from us. Gaborone is also *'tengnyanateng'*.
>
> (Mogwe, 1992, Introduction)

As this implies, Basarwa see their country as home, and the city as remote. That perception emphasises their strong attachment to traditional land and must cast doubt on the effectiveness of programs which aim to redress their socio-economic inequality by encouraging them to move to more accessible locations. Even the concentration of settlement within rural areas, a basic component of RADP, has caused them significant social disruption.

Assimilation and cultural differences

Culturally, Basarwa society, as writers such as Marshall (1976), Lee (1979) and Wilmsen (1989b) have documented, is that of a hunter-gatherer economy in an extremely harsh and arid environment where human survival depends on kinship and communal support systems. These features contrast markedly with societies which promote individual ownership of resources and entrepreneurial economic activities. Botswana's thrust for development has emphasised the latter attributes and government policies assume that all people will eventually be assimilated into that more individualistic society. Among other things that assimilation process has involved the promotion of common language, school curricula and forms of local government. In the process many Basarwa have experienced significant loss of culture and identity.

Land and resources

Basarwa once occupied the whole of Botswana. Although for over a thousand years other settlers, such as Bantu tribespeople, Afrikaaners and British have arrived and taken up residence the fact of that original occupation remains important. Under Basarwa traditional land ownership the country was loosely demarcated into territories (*n!ore*) with sufficient economic resources for the group concerned (e.g. Lee, 1972; Wilmsen, 1989a). That demarcation left little visible sign on the landscape. As incoming pastoralists and their herds gradually encroached on the country they intermingled with Basarwa and displaced them from their hunting grounds. Basarwa sometimes retaliated by killing cattle (Hitchcock, 1987) but ultimately they had little choice. Most opted to work for the herders in exchange for goods such as clothing and food. The loss of their territory, and the resultant loss of the vital resources for hunting and gathering, forced them into almost total dependency and, according to Biesele *et al.* (1989), has been the chief cause of their marginalisation . Other elements of Basarwa land tenure of which the incomers were ignorant included the group nature of their ownership and the vital role of kinship structures which determine that group membership; and how, through marriage, people have overlapping interests in different *n!ore*. This ignorance has subsequently created problems in applying conventional land tenure law, which is largely based on British interpretations that land must be individually owned.

Today eastern Botswana, where the land has for long been almost totally alienated for pastoral and urban development, contains less than 10 per cent of the Basarwa population (Biesele *et al.*, 1989). In western areas such as the Ghanzi region, where the land was sub-divided for leasehold farm development over a century ago, many Basarwa became pastoral and agricultural workers. Only in more remote areas such as the central Kalahari were Basarwa groups able to maintain their traditional ways of life until relatively recent times. In the 1960s continuing pressures on these remaining 'untouched' areas led to recommendations for their declaration as reserves so that the land and water resources and the remaining Basarwa hunter-gatherers would have some protection (Silberbauer, 1965). As the eventual declaration of a large region as the Central Kalahari Game Reserve (CKGR) suggests, the necessity for that protection was perceived not primarily in terms of human needs but more in terms of wildlife conservation. Although the Botswana government has since established special programs to counteract some of the disadvantages resulting from land and resource loss this is still the major factor underlying contemporary Basarwa problems.

Botswana's development aims and Basarwa interests

Botswana's new-found wealth has been based on two main forms of resource development: diamonds, which accounted for 80 per cent of export earnings in 1990 (Good, 1992); and commercial cattle production, associated with favourable marketing contracts supporting the export of meat to the European Community. Tourism, although still of only minor importance, is also now being encouraged. All of these forms of development have benefited from high levels of national and foreign investment and depend heavily on external markets for their profits. Their operations also impinge on Basarwa interests.

Diamond mining is currently concentrated in eastern Botswana sites. However prospecting has now spread through the interior of the CKGR and pressures to remove the remaining Basarwa hunter-gatherer groups from these areas to centralised settlements have been stepped up. Commercial pastoralism rests on modern technology, such as fencing and the sinking of boreholes for water. New techniques have enabled both miners and pastoralists to tap water at greater and greater depths, thus penetrating the Kalahari sandveld and spreading cattle further and further into the desert. Although current debate (Perkins and Thomas, 1993) suggests that cattle grazing may ultimately be highly destructive of this very fragile environment, Botswana's perception of cattle rearing as the basis of human wealth is likely to encourage these practices even more. Tourism also threatens Basarwa interests in the Kalahari. Basarwa are perceived as threatening wildlife, seen as the major tourist attraction in the desert. While Basarwa families undoubtedly harvest wild meat when possible (Kann et al., 1990) the quantities obtained today cannot be large. Restrictions on Basarwa hunting, including suggesting that they should only be allowed to do so with traditional poisoned arrows and spears and severe laws regulating firearm ownership, have had a profound effect. Indeed, as Mogwe's (1992) study demonstrates, high rates of Basarwa imprisonment can often be attributed to flouting of these regulations. Because Basarwa lack the money to pay fines detention, sometimes illegal, is almost inevitable and important human rights questions arise.

What can these forms of development offer Basarwa? Although some have, like many other rural poor in Botswana, periodically worked in South Africa's mines to earn cash, job opportunities for unskilled people in modern mines such as that operated by Debswana Diamonds are limited. They lack the vital resources – land, water and stock – to gain a toehold in the commercial cattle industry. And, although tourism may offer opportunities for Basarwa arts and crafts producers and for people willing to work as safari guides (Hitchcock and Brandenburgh, 1990), these chances are only slowly being realised.

Government measures to redress socio-economic inequity

Although Botswana has resolutely followed a non-racial policy designed to avoid the pitfalls of separate development the needs of disadvantaged sections of the population have not been ignored. The situation of rural dwellers in general, and the plight of landless families in particular has been of major concern. Clearly many Basarwa families fall into these categories. Government measures designed to assist these groups, however, assume that their needs and aspirations will accord with those pervading urban and land-holding Batswana society. These measures are therefore strongly assimilationist.

The Remote Area Development Program

Botswana's main scheme for assisting rural dwellers is the Remote Area Development Program, officially established in 1978. While this scheme caters for all Remote Area Dwellers (RADs) it owes its origin to an earlier program designed specifically to help Basarwa. By 1974 it had become increasingly obvious that Basarwa in western Botswana, particularly those whose traditional country lay in the vicinity of Ghanzi (Figure 1.2) were being traumatically affected by development. Many of them were already landless. Moreover because they neither ploughed nor raised cattle they had no chance of making successful applications for land grants (Wily, 1978). In addition, the introduction of more capital intensive forms of production on the Ghanzi farms had decreased demand for their labour and had transformed many of them into squatters, increasingly in conflict with the farmers because they stole cattle to keep themselves alive. Basarwa from further east in the CKGR had also become squatters in the settled areas, drawn there in search of water during the 1960s drought. The government's 1974 Bushman Development Program (BDP) was designed to deal with these problems by providing these Basarwa with stable communities. It was based on the establishment of centralised settlements in areas of land sufficient to provide some element of self-sufficiency through cultivation or ranching. As Wily (1982) points out it was then politically acceptable to designate this scheme as assistance for Bushman Development because it aimed to solve problems caused by the earlier British colonial regime rather than by the newly independent Botswana government.

By 1978 such views were no longer accepted and the program was renamed the Remote Area Development Program (RADP). It targeted not only Basarwa but all those who lived outside village settlements, that is, who were not part of conventional centralised communities with local councils controlled by a chief/headman. The characteristics of the RADs at whom this program was aimed included living far from basic services, having inadequate incomes, having inadequate or no access to land or water rights and being

21

culturally distinct. As Saugestad (1993c) rightly comments such attributes indicate a wealth of social problems. A brief description of contemporary conditions in a typical RAD settlement follows.

'Kweneng'[2]: A Remote Area Development (RAD) settlement in the eastern Kalahari (from Young, Fieldnotes, 1993)

'Kweneng' lies at the end of a dusty three-hour drive from the capital, Gaborone. Beyond stretches the infinite horizon of the Kalahari, an arid landscape of scrubby acacia thorn bush, small trees and limited water sources which in earlier times provided the whole resource base for the Bushmen traditional owners. Today, instead of providing a home for antelopes and predators, that country is cattle territory and the Bushmen are the herders, responsible to the absentee cattle owners for the day-to-day welfare of the stock. They live in the small mudbrick houses which spread around 'Kweneng's' concrete core – the newly built clinic, office and headman's complex; the school and teachers' houses form another large new compound on the northern edge of the settlement.

The three hundred or so Basarwa who live in 'Kweneng' are numerically very much in the majority but they lack political and economic power. They own virtually nothing and, while some hold jobs on government-funded projects such as the garden or spend time away working on cattle posts elsewhere, most depend for their livelihood on drought relief. This provides only a minimal income – in 1993, 98 pula (about $US 35) per month – and, because it is dependent on the severity of seasonal droughts, is also unpredictable. Drought relief workers carry out a wide variety of labouring jobs around the settlement – road repairs, general cleaning and mudbrick making for building construction.

Although there are only a few non-Basarwa families they hold important positions in the community. They own most of the village's livestock; they occupy most of the more permanent jobs, both in the village and in the school and the clinic, and run small businesses such as trade stores. Powerful leaders such as the village chief and the primary school headmaster are non-Basarwa who have moved to the local community from elsewhere.

Communication between non-Basarwa and Basarwa, while friendly enough, usually has paternalistic overtones. People know their places in the social sphere. In the school the teachers diligently instruct their pupils in Setswana and English and appear unaware even of the name of the local Basarwa language; in the clinic Basarwa mothers are similarly instructed on the care and nutrition of their children and may be chastised if they fail to take advice; and public discussions in the *kgotla*, the village meeting place, reveal that important decisions have often been made without consultation with those most affected, the Basarwa. Thus people express no surprise when officials of the local land board arrive unannounced to survey a site for a bottle shop which they had not asked for. The applicant is a non-local, with good political connections and a healthy awareness of the potentially lucrative market which he will tap. Most of his customers will inevitably be Basarwa. And while they may not complain about the new liquor shop it will certainly affect their families – scarce cash resources needed for purchase of food and clothing will be syphoned off and

22

there will be increased social disruption. Older Basarwa feel that the old ways, life on the edge of the Kalahari with none of these so-called amenities, were better. The young, who have grown up in 'Kweneng' and who have had little opportunity to learn of either the hardships or rewarding experiences of former times, are probably less convinced. The strong attraction of the bright lights of Molepolole or Gaborone may eventually pull them in the opposite direction, irrevocably further undermining Basarwa cultural autonomy in the process.

As this description of 'Kweneng' suggests the RADP, through its concentration on social welfare and services, has certainly helped disadvantaged people in small remote communities to obtain better water supplies, schools and clinics. It is estimated that around 51,000 Batswana currently participate in the program. Overseas aid agencies such as the Norwegian Agency for Development and Cooperation (NORAD) have provided generous assistance. But social service provision is only one part of the story. The RADP is also supposed to foster lasting economic development in remote areas. Here its achievements have been less impressive. As Kann *et al.* (1990: ix) stress, in 'economic development and employment, the provision of land rights, and education and training beyond a few years at primary school ... achievements ... are far less than can be observed in infrastructure development'. Post primary education is an essential element in self-determination. Without it people will not be able to take over skilled government funded

Plate 1.1 'Kweneng' schoolchildren

23

Plate 1.2 RAD vegetable garden project, 'Kweneng'

jobs in their own communities, and the present situation of dominance of outsiders in such positions, as clearly occurs in 'Kweneng', will persist.

A key element of economic development for RADP clients must be land rights. By gaining control over land and resources, people obtain the stability essential for the planning of their future. RADP failure to provide effective economic development reflects both the intransigent problems of finding suitable projects in remote areas and also the inappropriate nature of the approaches used. The list of RAD economic development projects, described in earlier studies such as Childers *et al.* (1982) and part of a contemporary community such as 'Kweneng', includes livestock projects, gardens and nurseries, chicken raising and bakeries. All of these should help to foster economic self-sufficiency. However their effectiveness seems to have been undermined by some chronic difficulties – most were initiated by outsiders rather than the people themselves, the RADs, and hence they may not be wholeheartedly supported. Outsiders are the supervisors and organisers and little responsibility for day-to-day operations is passed over to RADs. Also, because these projects do not generate many jobs, many people are still forced to depend on drought relief for their cash income, or must move elsewhere in search of work. Applicants to the RADP's Economic Promotion Fund, set up to help people wishing to set up their own enterprises, have to profess a commitment to economic rather than socio-economic development if they are to receive support. This may be difficult, and indeed

24

culturally inappropriate for Basarwa who still view their kinship networks and social interdependence as essential bases for support. Another problem, occurring in 'Kweneng' and common elsewhere, concerns competition for the use of the infrastructure in RAD settlements. Despite the fact that the government has funded RAD water boreholes specifically to meet the needs of the local residents and their stock, other people have often encroached and have established themselves in the area specifically to take advantage of such services. Effective control of such practices, and indeed self-determination in RAD settlements, is unlikely to occur under existing systems of administration where headmen are drawn from outside the community and where Basarwa in particular exercise so little political clout.

Tribal Grazing Land Policy (TGLP)

In 1975 the Botswana government initiated the Tribal Grazing Land Policy (TGLP), a plan to improve pastoral land management while at the same time instituting measures to assist landless people. This policy aimed to bring about greater equality in rural incomes and also enhance Botswana's earnings through a more commercialised livestock industry (Hitchcock, 1980). These two aims have proved hard to reconcile. Under TGLP land was to be zoned into three main categories: land for commercial use, which would obviously help in making pastoralism more commercially viable, and land classified either for communal use, or reserve land – categories within which it should be possible to assist landless people. In practice applications for commercial use land have generally taken precedence over all others. Hence TGLP seems to have assisted national economic development goals, and the interests of a few large-scale pastoralists, at the expense of meeting the needs of the rural poor. Good (1993) comments that in 1990 over 18,000 cattle farms, almost one-third of the national total, held an average herd of only 6 beasts while 35 commercial farms (0.06 per cent of all holdings) averaged over 4,100 head each. According to White (1993), 70 per cent of Botswana's people have no cattle, an astonishing situation in a country where cattle ownership remains the prime criterion of wealth and status.

For Basarwa the key elements of TGLP were communal and reserve lands, within which they would be free to follow their own lifestyle priorities, often combining hunting and gathering with other income earning ventures, including small-scale pastoralism. These needs were largely ignored. As Hitchcock and Bixler (1991) explain, many final zoning plans for land use under TGLP included only commercial and communal areas. TGLP effectively excluded reserve land applications because no land remained for these purposes. Later designation of some regions as Wildlife Management Areas (WMAs) has perhaps increased the chances of Basarwa who live in those areas having access to land for hunting and gathering purposes, but since these activities are also heavily scrutinised by wildlife rangers the possibilities for conflict are significant. Also, as Hitchcock and Holm (1991) point out, the chances of Basarwa in RAD settlements qualifying as communal land holders are remote because the amount of land held by such settlements is probably too small to be viable. Altogether it is somewhat ironic that, as White (1993) shows, the push for commercial pastoralism may have been quite misguided. Not only might it be unsound in terms of sustainable development and the environment, but it ignores the fact that the vast majority of cattleholders in Botswana practise subsistence pastoralism, for the day-to-day sustenance of their families as well as for cash.

Alternative approaches to Basarwa development?

Neither the RADP nor the changes brought about through land redistribution under TGLP have had radical effects on Basarwa socio-economic disadvantage; nor have they assisted the process of self-determination. What then are the alternatives? In Botswana itself programs introduced by the Kuru Development Trust (KDT) may contain lessons for achieving some progress. And in neighbouring Namibia the Nyae Nyae Development Foundation provides important examples of appropriate development on a larger scale.

D'Kar: the Kuru Development Trust (from Young, Fieldnotes, 1993)

D'Kar, a settlement of 300 to 400 people, lies on its own 3,000-hectare farm in the Ghanzi blocks on the main road between Ghanzi and Maun (Figure 1.2). To its east stretches the Central Kalahari Game Reserve (CKGR), the country to which some D'Kar families, the majority of whom are Basarwa, trace their traditional origins. Most D'Kar Basarwa, however, are Nharo people who have for several decades lived on or close to the Afrikaaner owned farms in the surrounding region.

On the surface D'Kar looks much like other remote settlements in Botswana: the government primary school compound, larger than most because its centralised location makes it suitable for providing boarding facilities for more isolated children; the village store; the garage and petrol bowser; and the village of mudbrick houses, scattered around this central core. But on closer inspection it turns out to be very different. The core also includes a complex of solid log buildings, including a church and pre-school, a residential hostel, a leather workshop and an arts and crafts centre. Elsewhere there is a tannery, a vegetable garden and a silkscreen workshop. The newest project, highly unusual and hopefully lucrative, is producing cochineal from worms which feed on the leaves of a cultivated patch of prickly pear.

D'Kar also differs from other remote settlements in the extent to which Basarwa are both involved in and are taking control over social, economic and political aspects of their lives. It is the home base of a Basarwa indigenous organisation, the Kuru Development Trust (KDT), a body focused on helping the people to gain control over their own lives and grapple more effectively

26

with the challenges with which development confronts them. With the support of a small number of dedicated people associated with the Dutch Reformed Church, the KDT and its Board of Trustees has targeted major problems such as poverty, culture, training and education and economic support. It co-ordinates all the separate economic projects in D'Kar, administering their financial accounts and ensuring that their plans fit as far as possible into a coherent whole, a vision for the future. Basarwa are involved in all aspects of this work, providing the labour in the various workshops and also gaining the necessary experience to take control of management and planning. Younger, literate Basarwa men and women, dealing with the day-to-day complexities of the financial and other administration of the KDT, translate their knowledge and experience to their elders and together these groups try to ensure that the community makes the key decisions about its own future.

The maintenance of Basarwa cultural identity and integrity is a vital principle of development at D'Kar. Language and literacy training, centring on the recording and production of material in Nharo, the principal local Basarwa language, is a key focus. The KDT pre-school, where the youngest children are able to learn in their mother tongue from Nharo teachers and where parents and grandparents are heavily involved in cultural activities is the practical demonstration of this. KDT have not confined their pre-school activities to D'Kar. They run similar establishments in half a dozen other remote communities in the Ghanzi region. Kuru Arts and Crafts also provides an obvious focus for cultural maintenance. KDT not only provides a marketing outlet for Basarwa traditional craftspeople from D'Kar and from other more remote settlements in the Kalahari. It is also the centre for Kuru artists, a group who are beginning to make their mark both within Botswana and overseas. Inspired by what they saw on a visit to caves in the Tsodilo Hills, they have rapidly developed their own natural skills to produce paintings which are stunning both in design and colour. These have been successfully exhibited in Gaborone and more recently in London, where the presentation was made by a number of the artists themselves. At present the future for these artists seems bright.

KDT has not only allowed the people of D'Kar to learn how to promote self-help ideals through education, work experience and the establishment of a variety of activities with money earning potential. It has also promoted Basarwa cultural revival, through literacy training and work with the language pre-schools, through art and artefact projects and through other activities such as hunting and gathering trips to the Kalahari and cultural festivals. Experiences gained through initiating and organising such projects has also given them much greater confidence, both in their own abilities in running the community and in their ability to speak out about Basarwa issues. Evidence of this has come from their recent contributions to public debate, both in meetings specifically concerned with Basarwa issues (e.g. RADP seminar at Ghanzi, September 1992) and in conferences where Basarwa issues are only one component in the debate. At a 1992 Gaborone conference on Sustainable Rural Development Komtsha Komtsha, the chairperson of KDT, talked in public of three main problems of the Basarwa of the Ghanzi area – loss of land, alcoholism and the seduction of young Basarwa women by Batswana men. This was the first time that a Nharo person had made such statements in such a venue. This resulted in a further meeting for discussion of land issues between Basarwa, government officials and politicians. Unfortunately in this forum statements about land aspirations and Basarwa self-determination were construed as confrontational and misguidedly reported by the media as amounting to demands for secession

27

and/or separate development. Nevertheless the Botswana government supported a regional San/Basarwa conference in late 1993 and useful dialogue between Basarwa and the bureaucracy appears to be opening up. People from KDT are prominent in these efforts, not only in the public arena but also in consulting with fellow Basarwa from other settlements in the Ghanzi area so that they can establish their own 'First Peoples' organisation. As Saugestad (1993c) comments, such indigenous bodies are necessary if effective consultation and negotiation with government is to take place.

Nyae Nyae

Namibia provides an important example of an alternative form of development not yet attempted in Botswana. The Nyae Nyae Farmers Cooperative (NNFC) and Nyae Nyae Development Foundation (NNDFN) are complementary organisations representing the Ju/'hoan people of Eastern Bushmanland or Nyae Nyae, a part of Namibia which adjoins the northwest part of Botswana. Their establishment stemmed from determination on the part of the Ju/'hoan to reoccupy the remaining part of their traditional lands (about 6,000 sq. km out of an original area of about 50,000 sq. km) to prevent any further alienation of their country. Escape from the marginalised squatter

Plate 1.3 Basarwa village at D'Kar, Botswana

28

Plate 1.4 Kuru artists' workshop, D'Kar

Plate 1.5 D'Kar community store, which not only sells food and drink but also buys and sells Basarwa arts and crafts from surrounding communities

camps around the small town of Tjum!kui, to which dispirited and dispossessed groups of Ju/'hoan had gravitated during the 1960s and 1970s, was the other major incentive. Today NNFC, begun as a grassroots organisation in 1986, has over 1,000 Ju/'hoan members living in 35 'outstation' communities in Nyae Nyae (Biesele *et al.*, 1991; NNDFN and NNFC, 1992; Hitchcock, 1992b). Its contribution so far has been in education (based on Ju/'hoan literacy and bilingual teaching), training in a wide variety of skills useful to support the mixed subsistence farming which is the economic basis of these groups, and communications; and, above all, in securing recognition from the Namibian government that the Ju/'hoan are the traditional owners of the Nyae Nyae lands. In the latter case appropriate land use planning, including wildlife management, has been an important outcome. NNDFN, which began as the Ju/wa (Ju/'hoan) Bushman Development Foundation in 1981, is a Windhoek based support NGO involved in developing training and support for economic development; for self-sufficiency; and to assist Ju/'hoan people in acquiring the organisational and communication skills needed to negotiate for their place in Namibia.

Significant differences between NNFC and NNDFN on the one hand and KDT on the other include the land base; and the recognition of indigenous groups as representative bodies with which government is willing to negotiate. Although Nyae Nyae is small in relation to the original territory of the Ju/'hoan it is still, at 6,000 sq. km for about 1,100 people, huge compared to the 30 sq. km held by the 400 or so residents of D'Kar. Moreover, despite problems in negotiation with government (see Hitchcock and Bixler, 1991) the Namibian government has to some extent been willing to listen to arguments presenting the cultural characteristics of Ju/'hoan land tenure, and presumably accept the importance of these. However the main difference is that NNFC has been accepted by the Namibian government as a representative grassroots organisation to whom they can talk. This stage is yet to be reached in Botswana but the KDT at D'Kar may well offer the first opportunity for the establishment of such a body.

CONCLUSION: RIDDLES IN ABORIGINAL DEVELOPMENT

As the example of Botswana's Basarwa people demonstrates, aboriginal development processes are affected by many opposing forces and attempts to balance these pose questions which are hard to answer. These forces include government promotion of top-down development compared to the people's desires for development which grows from their own community-based perceptions; feelings that cultural and social differences should be erased because they stand in the way of reaching the goal of assimilation to a national or regional identity; the perception that the national good should override local aspirations and needs, even when it is destructive of the latter; the encouragement of development in growth areas, usually urban, and the resultant marginalisation of remote rural peoples; and the increasing socio-economic disparities which stem from such uneven development. These themes are not peculiar to an indigenous minority group in a third world country which is itself grappling with development dilemmas. They also emerge strongly in examination of the development situation of remote-dwelling indigenous minorities in industrialised countries, the 'first peoples' of Australia and Canada. There they could be appropriately described as forming a 'third world in the first'.

NOTES

1   Commonly used terms for the known pre-colonial inhabitants of a country are 'aboriginal peoples'; 'indigenous (minority) peoples' (depending on whether or not they form a minority in the population); and, increasingly popular, 'first peoples'. In North America 'native peoples' was until recently commonly used as a collective term but in Canada this is now often replaced by 'aboriginal people'. In this text I intermingle 'aboriginal' people, 'indigenous (minority)' people and 'first people'. When specific Canadian groups are referred to I use Indian, Inuit and Metis. With Australian groups I use Aborigines and Torres Strait Islanders.
2   This description is based on a short tourist visit to a settlement in the Kweneng district when I accompanied a journalist reporting on the progress of the RAD program. It has been extended to include traits which, according to other reports, are common to RAD settlements and I have therefore given it a fictitious name – 'Kweneng'.

31

2

REMOTE AREA DEVELOPMENT IN AUSTRALIA AND CANADA
Perceptions, people and resources

Remote areas, the 'outback' or the 'wild', are important to Australians and Canadians. High levels of urbanisation and population concentration inevitably mean that for the majority of the people in these countries large parts of their region are in fact remote, far from where they live and from the hub of economic, social and political life. Geographical definitions of those remote areas have, in the case of Australia, encompassed the tropical lands (Courtenay, 1982: 42–4), including the wetland and rainforest areas of the monsoonal 'Top End' and Cape York, the vast savanna regions of the semi-arid zone and Gulf country and the desert, stretching from the West Australian coast east to beyond the Queensland border and south into South Australia. Together these areas cover almost half the Australian continent. In Canada remote areas, defined as the Arctic and sub-Arctic regions, including the whole of Yukon and Northwest Territories and extending to the southern edge of the boreal forest in the Prairie provinces, Ontario and Quebec have been estimated to account for almost 75 per cent of the country (Bone, 1992: 2–5). Geographical remoteness, characterised by difficult communications and resultant high costs of transport for both goods and people, is coupled with cultural distinctiveness, with Canadian and Australian 'first peoples' forming a very high proportion of the populations of these regions. It is on these remote areas, loosely defined according to the above criteria, that this study of aboriginal development focuses.

Development in remote parts of Australia and Canada has clearly operated in parallel with the processes discussed in the broader global context. It has been governed by resource exploitation, within a colonial ethos promoted by the economic cores in southern and eastern parts of these countries; diffusion of gains through this 'top-down' approach to development has been limited and many living in remote areas would perceive that their costs, in terms of environmental and social degradation, outweigh the benefits; and recent decades have seen increasing commitment to alternative forms of development, including community involvement and an emphasis on sustainability in all facets of society. These regions as many (for example, Ironside, 1988) have commented, are true marginal lands.

Aboriginal peoples form a distinct group within these areas. Their differences in social, cultural and economic terms, coupled with the paternalistic nature of the policies which have dominated their administration for decades, have led to them being referred to as the 'Fourth World' (Manuel and Posluns, 1974; Dyck, 1985). As writers such as Weissling (1989) have suggested, their experiences of development demonstrate some analogies with those of their 'third world' peers. They have been variously described as experiencing under-development (Pretes, 1988a); marginalisation (Chamberlin, 1983; Drakakis-Smith, 1980; Dagmar, 1982); and colonialism, both internal and welfare (Paine, 1977; Hartwig, 1978; Drakakis-Smith, 1984b; Frideres, 1983; Ponting, 1986).

PERCEPTIONS OF DEVELOPMENT

Colonialism and resource exploitation

Non-aboriginal pioneers travelling to remote parts of Canada and Australia saw the resources of the land primarily in terms of commercial exploitation. Thus Alexander Forrest gives the following description of his first impressions of the grasslands of the remote east Kimberley area of Western Australia in the early 1880s:

> The land ... to the east would be a cattle man's paradise with vast plains of splendid pasture, much of it well watered and not subject to flooding since the land is higher and the river contained between deep banks.
>
> (Forrest, quoted in Durack, 1959)

Not surprisingly, southern graziers, led by the Queensland-based Durack family, rose rapidly to this challenge and within a remarkably short time their herds of cattle were spread over most of the accessible lands and beasts were being shipped south out of the rapidly growing port of Wyndham. The Aboriginal people of the region were dispossessed of their lands, many were massacred or died from the ravages of introduced diseases in the process and those who survived became the backbone of a richly rewarding pastoral industry. They were not always harshly treated. As Mary Durack's biographies of her family demonstrate (Durack, 1959; 1983), and Berndt and Berndt (1987) and McGrath (1987) confirm, that relationship was frequently one of mutual respect and support. Nevertheless it was exploitative. Aboriginal stockworkers did not gain access to award wages until the late 1960s and many today still recall the substandard living conditions to which they and their families were subjected in the recent past. This colonial form of development, sometimes referred to as an example of internal colonialism (Hartwig, 1978), bears some similarity to the processes experienced under the imperial systems of Britain in Africa or the Dutch in Indonesia.

The early development of the Canadian north provides other examples which, while different in detail, had similar effects. Here the attractive resource, fur-bearing animals, led to development under the auspices of the Hudson's Bay Company which established its commercial empire in Rupert's Land in the mid-seventeenth century. Aboriginal people, with their detailed knowledge of the environment and its wildlife, provided the vital core of the labour force – the trappers whose products were then sold through middlemen traders to the company. In contrast to pastoralism in the Australian rangelands, this did not result in wholesale resettlement of aboriginal peoples from their traditional country. However, as Ray (1984) shows, it did tie them into a colonial relationship. Their efforts were rewarded solely in kind, with the company operating for many years on a currency called the 'Beaver', after the main product which it purchased; and they were forced to give the company their trade because they could not escape from the system of debts and credits which provided them with their living.

While colonial aspirations arguably governed aboriginal development through both the pastoral and trapping industry, other types of remote area resource development could be conducted without the involvement of local residents. The early mining industry, in the form of successive 'rushes' to the goldfields of Canada's Yukon or Australia's Northern Territory and Western Australia, was a haphazard affair, dominated by outsiders intent on getting rich quick. The 'boom and bust' mentality which this entailed was, however, maintained in the modern mining industry which has continued to emphasise large-scale operations, externally controlled, owned, managed and financed and with only limited benefit to local residents. Until recent times aboriginal peoples have not only had little to do with this industry but they have arguably been marginalised by it. Transformation of this situation caused by the recognition of aboriginal land rights and concern over the detrimental environmental effects of such operations has only recently begun to have an effect.

These types of development were all implemented from the 'top-down', under external control. Not only was there no direct consultation with aboriginal people affected by these processes but, unless their labour contribution was essential, they were completely excluded. Their continued presence in these regions was tolerated, but it was anticipated that, through time, they would disappear as a distinct cultural group. This would occur in two main ways. Some aboriginal people would leave the land for ever and move to the cities in search of jobs and the good things in life. The remainder would eventually die out. The latter assumption was based on observation of their appalling health status and obviously declining populations, and resulted in a policy of support for the status quo, sometimes called, in Australia, the 'smooth the dying pillow policy'. That policy assumed that survivors should as far as possible continue to follow their traditional economic activities in living off the land. As a result government authorities made little effort to provide services or development opportunities for aboriginal people in remote areas. That responsibility was largely taken up by the missionaries.

Assimilation versus self-determination

The Second World War, in both Australia and Canada, was a turning point in remote area development. As Abele (1987) describes, the strategic importance of northern Canada both during and immediately after the war resulted in the construction of roads and oil pipelines and military installations. Increased contact between those who migrated north to work in these new industries and the local population revealed the economic and social hardships being faced by aboriginal peoples. This resulted in much more determined efforts to provide administration and services under government auspices. Assimilation, whereby aboriginal peoples would discard their former ways in favour of those essential to the industrial way of life, became the cornerstone of the new policy.

Australia's experiences in this period were similar. Here Aboriginal involvement in the construction of all-weather roads for strategic and development purposes, particularly the Stuart Highway linking the Northern Territory to the remainder of the country, created a new form of contact with non-Aboriginal society. During the months which Aborigines from remote parts of the desert spent as labourers on the road they learned new skills, earned cash wages and for the first time saw dark-skinned people, here American soldiers, being treated as equals. Not surprisingly many were later unwilling to return to their old way of life in the desert and the growth of large mission and government administered settlements was a natural consequence. As many have documented (for example, Rowley, 1970a and 1970b; Long, 1970; Smith, 1980b; Stanley, 1985), it was in these 'new towns' that assimilation policies began to have their full effect on remote-dwelling Aborigines – their children were sent to school, their diseases were treated in government clinics, they were encouraged to live in houses and adopt modes of non-aboriginal behaviour and, at all times, they were instructed in the value of the non-aboriginal work ethic. Altogether, in both the Canadian and Australian situations, assimilation was seen as the vehicle for the successful implementation of externally promoted forms of development.

Even in its early stages the success of the assimilation policy in remote regions was quite limited. Coates (1988) describes how, despite educational curricula specifically designed to exclude Indian/Inuit cultural knowledge and language practice, aboriginal people in Canada's Yukon Territory in the 1950s held on to these vital components of knowledge. Similarly, although school attendance was enforced partly to bring people into the larger centralised settlements, many Yukon people still continued to live in more remote, smaller communities. The response in outback Australia has been comparable; Aboriginal school children adapted themselves to a dual system – a school day spent communicating in English and learning about Australia's 'discovery' by Captain Cook, and the remainder of their time spent within an Aboriginal cultural milieu, in which English usage and many aspects of non-aboriginal lifestyles were of very little consequence. Today governments in both countries recognise the strength which lies behind this survival of aboriginal culture and no longer overtly follow policies aimed at full assimilation.

From the 1960s onwards self-determination, whereby people themselves have the right to determine their own lifestyle and take control of their own affairs, replaced assimilation as the basis for aboriginal policy in both Canada and Australia. Self-determination would, it was assumed, be accompanied by self-management, a concentration on self-sufficiency, the concept of self-government and a community-based development approach. It was no accident that these changes in direction coincided with processes of decolonisation and determination to achieve independence throughout the 'third world'. Such policy changes had some significant effects, such as the implementation of land rights legislation in both countries, the acceptance of the value of Aboriginal, Indian and Inuit cultural and social attributes in their own communities and the devolution of centralised bureaucratic control from Ottawa and Canberra to the 'bush'. They also, in development terms, were marked by new approaches which were much more strongly founded on community involvement and on consultation and which focused on the goals which aboriginal people wish to achieve. These changes came about largely because of pressures exerted by aboriginal peoples who, determined to break out of the assimilatory mould into which they were being forced, have increasingly demanded control over their own affairs.

Policies may have changed. What about the practical realities? As aboriginal people are well aware, many Canadians and Australians still favour assimilation, and are unwilling to condone government support for services and facilities planned to take account of aboriginal cultural concepts or aspirations. This has, in Australia at least, been described as a form of apartheid and an attitude that unduly favours aboriginal people. Federal government departments have also found it difficult to discard assimilation because the implementation of this policy was so well entrenched in their bureaucracies. Consequently the road to self-determination has been slow, tortuous and frustrating (see, for example, Driben and Trudeau, 1983; Bear *et al.*, 1984).

This problem, as was clear from discussions at an Edinburgh University workshop on this issue in 1986 (Macartney, 1988), is not restricted to Canadian and Australian aboriginal policy but has also been apparent in other Commonwealth countries and no doubt in other ex-colonial parts of the globe. Current economic stringencies have exacerbated the difficulty. Pressure to ensure that government funds are spent as effectively as possible, and that all funding is fully accounted for is at present particularly strong. It has resulted in calls for increased control over expenditure, and in the process self-determination must be undermined. Altogether, while the actual practice of self-determination or self-management is accepted, its implementation is continually hindered by the assimilationist tendencies of the system. It is hardly surprising that aboriginal people are frustrated. They are also confused, both by the continual policy changes and by the bewildering range of inconsistently defined terms with which they are bombarded. Thus Canada has been stressing devolution of bureaucratic centralisation through aboriginal community development, but people are uncertain what this means. Is it a holistic form of development, incorporating both economic and social aspects of the community; or does it apply primarily to economic development? And does it really come under community control, or is it merely yet another imposed form of administration, devised and implemented primarily under government supervision (Dacks, 1983)? Reluctance by Indian bands to enter into community government agreements with the Federal government of Canada suggests that many people suspect that the latter is the case (Wolfe, 1989). Similarly, in Australia's Northern Territory, proposals to set up community government in all centralised townships, many of which are primarily Aboriginal both in character and population, have not been wholeheartedly accepted and, although the scheme was introduced in 1985, only a comparatively small number of such schemes have as yet been finalised (Mowbray, 1986; Ellanna et al., 1988).

More recently attention has increasingly focused on the establishment of regional agreements by which the national government devolves responsibility for key components such as land and resource management, service delivery and economic development to the aboriginal people in the region concerned. Appropriate funding is negotiated and allocated. In Canada elements of such agreements have been included in land claim settlements for the Inuvialuit people of the Mackenzie Delta region and the Inuit people of Nunavut, here concerned specifically with land-use decision-making and wildlife management. Regional Aboriginal organisations in Australia, notably the Kimberley Land Council in northern Western Australia, are currently pushing for similar recognition of their needs to control their own country. Findings from a recently completed study of the Kimberley Aboriginal economy (Crough and Christopherson, 1993b) provide strong backing for such proposals by showing that government spending on Aboriginal needs and Aboriginal spending in the Kimberley currently amounts to about $140 million per year. If profits from the Argyle Diamond mine, which makes little direct contribution to the Kimberley economy, are excluded this accounts for about 40 per cent of the Kimberley Gross Domestic Product. With that sort of input the idea of Aboriginal controlled regional development in the area becomes quite practical.

Development policies for aboriginal peoples in remote parts of Australia and Canada have therefore had a chequered history which, although seemingly becoming more sympathetic to aboriginal needs through time, is still dominated by non-aboriginal ethnocentric thinking. Apart from the perpetuation of colonial attitudes, a major failure has been an unwillingness to recognise: firstly, that aboriginal people's perceptions about both the process of development itself, and about the ends it sets out to achieve are often different from those of non-aboriginal people; and secondly that, as in any society, there is a wide range of opinions which will be difficult to accommodate within a single rigidly structured policy framework. Remote area mining development offers a wealth of good examples. Inuit people who find themselves confronted with a large-scale, high-tech mining development which offers them no employment opportunities and gives them no share in the profits may well resist the project; however Dene Indians who have the chance of becoming joint venturers in a small mining operation in the Mackenzie Valley may well support mining. The Australian outback has produced similar contrasting attitudes, not always between distinct population groups but even within one group; the coincidence of mine sites and country of great spiritual significance has often been a major issue, with traditional owners of the country firmly stating that development is impossible and others refuting this argument.

Differences such as these can only be resolved through careful consultation both within the community, and between members of the community and the developers. Such consultation must occur within a holistic framework which incorporates social and economic elements. In stressing the key importance of that framework, it may also be useful to break it down into its separate components so that the relative importance of these can also be examined. Here a model proposed for northern Canadian development (Blishen *et al.*, 1979) offers a possible approach. This model is based on the belief that any form of development will have social, economic and political implications for any community, and that the community's response to these will determine how people feel about these changes. The model proposes examining three main elements within a community;

- *economic viability* – for example, the economic base of the community, its economic enterprises, the labour force, markets and incomes;
- *social vitality* – for example, the socio-demographic characteristics of its population and how the people are supported through health, education and welfare services;
- *political efficacy* – for example, the involvement of the community in local and regional government agencies and other types of community decision-making processes.

This model can be applied in many different types of community. It is certainly appropriate in the context of aboriginal development, where its holistic and culturally sensitive approach could be particularly useful.

Australian aboriginal development studies which make use of this framework include the East Kimberley Impact Assessment Project (EKIAP) (Coombs *et al.*, 1989) and a contemporary project on sustainable development and Aboriginal community planning (Young and Ross, 1993).

PEOPLE AND RESOURCES

Aboriginal peoples, like any other population groups, are not the passive recipients of development. They have their own ideas on what they would like and, increasingly, are free to choose the alternatives most suited to them. At the same time their choices are being extended because, with the vast improvement in modern communications systems, their knowledge of what is taking place elsewhere is much broader than in earlier times. These choices are affected by their social and cultural attributes, many of which contrast significantly with those common in non-aboriginal society. However any decision which they make may also be constricted by the realities of their socio-economic situation, particularly attributes such as their work skills, their incomes or their levels of education. Factors affecting such characteristics, including age structures or health status, also make an impact. Indicators such as these generally demonstrate that aboriginal people are severely disadvantaged compared to the rest of the Canadian or Australian populations.

Another crucial factor is their access to resources: ownership of the land and of its natural resources through forms of tenure which enable them to direct and control the development process; and access to financial resources which will give them the capital necessary to carry out their plans. All of these factors play a particularly important role in economic development for people living in remote communities.

Identification and definitions of aboriginality

As the Botswana example showed, the relative socio-economic status of different population groups can only be properly assessed if these groups are clearly identified in official censuses. Lack of such identification severely handicaps a study of the Basarwa situation. However, Canada and Australia now have such identification and in general in both of these countries it is the standard census definitions of aboriginality which are universally accepted. In general, the Canadian situation, where different aboriginal groups have historically been subjected to different treatment by governments and still have different rights of access to government support, is more complex than that in Australia. There, although Aborigines and Torres Strait Islanders form two distinct groups, many of whose members also claim a degree of non-aboriginal ancestry, the same overall policies have been applied at federal level .

39

Canada

Canadian native people are generally classified into three main categories: status and non-status Indians (now simply North American Indians), Inuit and Metis. People are allocated to these categories on the basis of self-identification, as illustrated by the question posed in the 1986 Canadian Census:

> To which ethnic or cultural group(s) do you or did your ancestors belong?

The broad term 'aboriginal Canadians' is used to describe all people who responded by classifying themselves as Inuit, North American Indian or Metis, any one who identified with more than one of these groups, and anyone who identified with any other ethnic group in addition to one of these. Since these three aboriginal groups have been differently treated by government authorities it is necessary to discuss them in greater detail.

The affairs of status Indians, those who 'are either registered, or are entitled to be registered under special federal legislation' (the Indian Act), (Printup, 1988: 60) are legally administered under the terms of the Indian Act, and are the responsibility of the federal Department of Indian Affairs, now called the Department of Indian and Northern Development (DIAND). Their reserve lands, which until the recent land claim era were the only lands over which they exerted some form of control, were held by DIAND on their behalf. The department's brief, somewhat paternalistically expressed, was to protect Indians against uncontrolled seizure of their lands and intrusion by settlers. This has had both positive and negative effects. As Beaver (1979) stresses, the DIAND Indian agent could control many aspects of life for status Indians resident on reserves, even to the extent of preventing them from moving elsewhere, or setting up their own businesses without permission. This control undermined people's independence and willingness to make decisions about their own lives. On the positive side the government also shielded these groups from some of the potentially disruptive activities which might have impinged on them from outside. Moreover the many programs provided by DIAND to promote social development, and the low interest loans which they could obtain to support economic enterprises, gave these groups a certain advantage compared to other native peoples.

Non-status Indians, both those who had lost their status through intermarriage with outsiders and the descendants of these unions, could not claim benefits from programs run by DIAND. Inevitably a large proportion of non-status Indians were women who had married non-natives, and it is hardly surprising that this feature of the Indian Act was increasingly criticised on the basis of sexual discrimination. After much discussion and deliberation the Act was finally amended in 1985 and subsequently many people (estimated at around 40,000) have applied for reinstatement. Assessment of their claims is based on evidence of family connections, plus recognition as a member of the relevant band. The new category North American Indian is now used to cover these formerly separated groups.

Metis people, those whose native Indian descent was mixed with Scots, Irish, English and French ancestry, have been less clearly recognised in official terms. Many families trace their Metis origins back to the days of the fur trade when they were often the middlemen between the Indian trappers and the Hudson's Bay Company traders. Their situation has been a classic case of assimilation which has failed. Although many Metis have become urban dwellers, and have adopted many facets of the industrialised economy, they have not wholly assimilated. They have retained a distinctive culture which incorporates both Indian and European attitudes and forms of behaviour – music, art and literature – and in recent decades they have increasingly demonstrated their group identity in ways which the government is no longer able to ignore. As part of this process they have asked to participate in some of the programs not previously available to them, and have also asked for authorities to establish special Metis programs.

Inuit were the most recent aboriginal group to come into direct contact with non-aboriginal Canadians. The extreme isolation of their Arctic homelands, coupled with the perception that these lands offered little of any commercial value to outsiders, protected many of them from large-scale intrusion until the present century. As a consequence they were excluded from the Royal Proclamation of 1763 which decreed that claims for alienation of land from native Canadian groups must take place through treaties of cession. Unlike the Indians, they signed no treaties with the incomers. They are not defined as Indians under the Indian Act itself. However, following a 1939 Supreme Court ruling on the Constitution Act of 1867, the federal government was judged to be responsible not only for 'Indians and lands reserved for Indians, but also for Inuit' (Printup, 1988: 61). DIAND has therefore administered special programs dealing with social and economic development in Inuit communities.

Australia

The contemporary Australian definition of aboriginality, covering Aborigines and Torres Strait Islanders, is also based on self-identification. The census question is as follows:

Is this person of Aboriginal or Torres Strait Islander descent?

It also states that a person claiming aboriginal origin must be recognised as such by other members of the community to which that individual belongs. This definition, unlike that used in Canada, is separate from other questions about ethnic origin. People are not asked to state whether they belong to any other ethnic groups, or what degree of aboriginal ancestry they claim.

Officially there are therefore no equivalents of the Metis in Australia. Special programs for aboriginal people, such as those administered by the Aboriginal and Torres Strait Islander Commission (ATSIC) for social and economic development, use the census definition to determine who will be eligible.

This contemporary definition of Australian aboriginality has only been legally recognised since the late 1960s. Prior to that time, as Rowley (1970a: 341–64) discusses, definitions had varied between states and territories and had contained a number of categories differentiated according to the degree of aboriginal ancestry. People were primarily distinguished according to whether they were of part Aboriginal descent, often referred to as 'half-castes'; or whether they were only of aboriginal ancestry, colloquially called 'full-bloods'. Clear distinction between these categories was impossible, although in general half-castes were assumed to have no more than 50 per cent of their ancestry derived from aboriginal parentage. 'Half-castes' and 'full-bloods' had different legal status, and were treated differently through the programs designed to promote the aboriginal policies of the day. Striking anomalies were common. The rules governing inclusion varied not only between programs but also between states/territories and it was quite possible for a single individual to find his or herself defined as an aboriginal person for one purpose, and simultaneously denied that status on another occasion.

Anomalies and complications such as these were partly responsible for the decision to hold a federal referendum in 1967 to determine both the future status of Australia's aboriginal population, and where responsibility for their affairs should lie. That referendum, passed by 91 per cent of the electorate who voted, gave the Commonwealth government the power to legislate on behalf of all Aboriginal people, overriding state responsibilities if deemed to be necessary. It also stated that all Aborigines should, in recognition of their status as citizens, be enumerated in the national census. The Australian constitution had previously excluded people of full Aboriginal descent from this provision. Such exclusion essentially denied them full rights of citizenship. It has also meant that complete census statistics for Australia's aboriginal population date only from the national count of 1971, when the above definition of self-identification was first used. As demographers have subsequently discussed (Smith, 1980a; Gray and Smith, 1983) the imprecise nature of the self-identification definition has continued to present many problems in the interpretation of census data. It is certain, for example, that many people did not identify their aboriginal ancestry in 1971; however they have subsequently identified themselves. This has made it impossible to measure the rate of aboriginal population growth by comparing total populations from successive censuses. Nevertheless, despite these problems, this contemporary definition is at least standardised and is much more acceptable to aboriginal people than the categories used in pre-referendum days. Unfortunately many other Australians have failed to accept identity as sufficient evidence of aboriginality and continue to comment disparagingly about aboriginal people who look like Europeans. The common assumption is that they are identifying their aboriginal origins for financial gain, an assessment which grossly overestimates the advantages offered to aboriginal people through government funding.

Aboriginal cultural and socio-economic characteristics

Although the Australian and Canadian censuses now similarly identify their 'first peoples', a useful breakdown of figures on an appropriate geographical basis is less easily obtained. It is therefore difficult to provide an overview of the basic demographic characteristics of remote dwelling aboriginal people compared to those in urban areas, and a more detailed analysis lies beyond the scope of this study. The following discussion therefore focuses on national comparisons. It is still valid because it highlights the general socio-economic disparities which affect the participation of all aboriginal groups in the development process. In 1986 the aboriginal populations of Australia and Canada were enumerated at only 1.5 per cent and 2.8 per cent of their national populations respectively (Table 2.1). In Canada the group was dominated by people of North American Indian origin (almost three-quarters of the total) and the Inuit accounted for less than 5 per cent. In Australia Aborigines, with almost 91 per cent of the aboriginal population, were by far the biggest group. As these figures demonstrate, numerically these 'first peoples' are now very much in the minority within their countries. They are however very distinctive. Two main points of difference occur – their social and cultural

Table 2.1 Australia and Canada: aboriginal population, 1986

	Nos	% total	% Aboriginal pop.
Australia			
Aboriginal	227,645	1.5	100.0
Non-Aboriginal	15,374,511	98.5	
Total	15,602,156	100.0	
Canada			
North American Indian	525,585		73.9
Inuit	33,465		4.7
Metis	128,545		18.1
Other	23,990		3.4
Total aboriginal	711,585	2.8	100.0
Non-aboriginal	24,310,420	97.2	
Total	25,022,005	100.0	

Source: Tables 2.1–2.7 are from Australian Census of Population and Housing, 1986; and Census, 1986, Statistics, Canada.

attributes, which identify them with ways of life largely foreign to members of the non-aboriginal society in Australia and Canada; and their socio-economic status, which is highly disadvantaged when compared to that commonly found in industrialised capitalist countries.

Aboriginal Canadians and Australians claim the right of prior occupation, as the 'first peoples' in these territories. Those claims are not under dispute, although in Australia the definition of the term occupation itself has been a source of disagreement since the earliest days of non-aboriginal settlement. As early chronicles document, many of the incomers, mostly of European origin, treated the land as if it was unoccupied. Not only were they unable to discern visible boundaries between the countries of different Aboriginal groups, but they were also unable to accept forms of land use which did not involve cultivation and which did not obviously transform the landscape. The mobility of Aboriginal people was also beyond their comprehension – they saw it as an 'idle and wandering habit' (Macquarie, 1816), not, as it would be more accurately described, as a process which was essential for the survival of a society dependent on a hunter-gatherer economy (Young and Doohan, 1989). These failures in cross-cultural understanding are an essential part of the background to the declaration of Australia as '*terra nullius*', a land that prior to European settlement belonged to no one. They have had a disastrous effect on Aboriginal people. They provided excuses for the seizure of their lands, and underlie the associated genocide and decimation of the population through disease. They also, as Reynolds (1987) discusses, explain the lack of treaties which would have provided a firmer basis for subsequent land rights. This contrasts with the situation in Canada, where the Royal Proclamation of 1763 recognised aboriginal title and laid the basis for treaty negotiations preceding settlement in most regions (Usher *et al.*, 1992). The Australian Supreme Court's Mabo decision of 1992, recognising native title for the Murray Islanders in the Torres Strait, at last legally overturns the principle of *terra nullius* and, with the recent passing of the federal government's Native Title legislation, has now given aboriginal people throughout the country the expectation that future land rights claims will be more fruitful.

These basic misunderstandings demonstrate that non-aboriginal settlers were largely ignorant about the nature of the relationship between aboriginal Australians and Canadians and the land. This fundamental relationship interweaves the value of the land for sustenance with its meaning in cultural terms. At a workshop held to discuss Aboriginal development issues in Australia's remote East Kimberley region participants expressed this simply as follows:

> Culture is a map. The land is the map. It is recorded on the land. . . . As we travel across the land, we follow the Law. Culture/Law tells us of our relationships to land and to our responsibilities to one another.
>
> (Kimberley Land Council, 1991)

And, in a statement made to the Mackenzie Valley Pipeline Inquiry in 1975, Rene Lamothe of Fort Simpson spoke as follows:

> The love of the Dene [Indians] for the land is in their tone of voice, a touch, the care for plants, the life of the people, and their knowledge that life as a people stems directly from the land.... [The land] is a storyteller, a listener, a traveller, ... a teacher, ... a benefactress.
>
> (Watkins, 1977: 11)

As these statements suggest, land is not merely a resource which provides the means for supporting human life. It is also the source of spirituality and cultural knowledge. In Australia that knowledge is recorded through stories of the ancestral origins, popularly called 'Dreamtime Stories'. These explain both the origins of the land itself – its landforms, its water sources, its plant and animal life – and also how people are related to specific areas within it. Oral evidence given in the course of land claim hearings (Young, 1992a), popular texts such as Chatwin's *The Songlines* (1987) and a host of academic studies have all brought knowledge of these relationships to the forefront in a way which was unknown to non-aboriginal Australians only a few decades ago. The interdependence of the aboriginal human–land relationship is so strong that breaking one part destroys the other. Thus when aboriginal people were turned off their land they lost many elements of their culture. Evidence from both Inuit and Indian communities in Canada shows the same effects and responses (see, for example, Brody, 1975 and 1981).

Aboriginal cultural attributes which affect economic development include kinship networks, reciprocal relationships, the strength of traditional authority structures and the relationships between different aboriginal groups. As I have discussed earlier in relation to Australia's Northern Territory (Young, 1988e), these forms of customary behaviour and social relationships affect all aspects of people's lives, including their economic activities. Thus people share resources with their kin, distributing not only cash and possessions through this network, but also giving kin favoured access to jobs or to positions of political importance; and these resources are shared with the understanding that, at the appropriate time, the share will be reciprocated. Such approaches must have been a fundamental means to survival under the traditional economy of hunter-gatherers, where many people suffered from unpredictable periodic shortages in food supplies, and where the welfare of the group was essentially founded on mutual trust. As can be imagined, they pose many problems when introduced into the management of conventional cash-based businesses and are a frequent focus of criticism from non-aboriginal bureaucrats charged with economic development support. Traditional authority structures, in particular the vesting of power for decision-making in the hands of those with the most profound traditional knowledge, also exert a strong influence. As those with that knowledge are almost invariably the older people this means that they have to be consulted

45

first about all aspects of development. Since, because of their lack of formal education, their knowledge of the non-aboriginal business world may well be less accurate than that of younger people, they may make decisions which inhibit development success. Yet progress is unlikely unless their views are sought. Finally, aboriginal people believe that each distinct group must speak for itself and for no other. This provides a healthy foundation for consensus decision-making, but makes it very difficult for representative groups – councils, boards of management – to operate effectively.

Issues such as these constantly arise in relation to discussions on Aboriginal development in other parts of Australia as well as the Northern Territory and, as Usher (1987b) and Wolfe (1989) have described, also influence the development process for aboriginal peoples in Canada. In both countries they are particularly important among population groups in remote areas, because it is in those regions that people have been able to retain the closest links to their ancestral lands, and where their original social and cultural structures have been strongly retained.

Geographic distribution

The geographical distributions of aboriginal populations in both Canada and Australia in 1986 were quite uneven. In Canada the largest numbers of aboriginal people lived in the provinces of Ontario, British Columbia and Alberta while the populations in the Atlantic provinces and northern Canada were comparatively small (Table 2.2a). In Australia, Queensland and New South Wales, with over half the population, had much larger numbers of aboriginal people than the states of Victoria and Tasmania (Table 2.2b). When aboriginal populations are examined in relation to their share of those of the province/state a different picture emerges. In both countries the most sparsely settled and remote areas, Australia's Northern Territory and Canada's Northwest and Yukon Territories, had the largest percentage of aboriginal peoples. However in the more populous southern regions, such as Victoria and New South Wales in Australia and Quebec and Ontario in Canada, they accounted for only small percentages of the total populations (Tables 2.2a and 2.2b).

Examination of the sub-groups of the aboriginal populations reveals further differences between states/provinces. While North American Indians were found in every province in Canada, their numbers were particularly large in Ontario, British Columbia, Alberta and Quebec. Metis, however, were concentrated in the Prairie Provinces of Alberta, Saskatchewan and Manitoba (66 per cent of the Metis population altogether); and over half the Inuit lived in northern Canada (Table 2.3). In Australia over 60 per cent of Torres Strait Islanders were in Queensland. Distributional differences such as these have important implications for development. Not only does the ethnic make-up, and hence possibly cultural attributes, of the aboriginal groups

Table 2.2a Distribution of aboriginal population, Canada, 1986

Province	Aboriginal population nos	% of total aboriginal population	Aboriginal pop. as % total provincial population
Atlantic[a]	34,345	4.8	1.5
Quebec	80,910	11.4	1.3
Ontario	167,380	23.5	1.9
Manitoba	85,235	12.0	8.1
Saskatchewan	77,650	10.9	7.8
Alberta	103,925	14.6	4.4
British Columbia	126,630	17.8	4.4
Northern Canada[b]	35,510	5.0	47.1
Total	711,585	100.0	2.8

Notes: [a]Newfoundland, Nova Scotia, New Brunswick, Prince Edward Island; [b]Yukon, Northwest Territories.

Table 2.2b Distribution of Aboriginal population, Australia, 1986

State	Aboriginal population nos	% of total Aboriginal population	Aboriginal pop. as % total state population
New South Wales	59,053	25.9	1.1
Victoria	12,569	5.5	0.3
Queensland	61,300	26.9	2.4
South Australia	14,289	6.3	1.1
Western Australia	37,768	16.6	2.7
Tasmania	6,719	3.0	1.5
Northern Territory	34,679	15.2	22.4
ACT	1,216	0.5	0.5
Total	227,593	100.0	1.5

vary between different areas but their colonial history has been different. Government policies continue to affect them differently.

The proportion of aboriginal peoples in rural and urban settings also varies. Both Canada and Australia, especially the latter, are highly urbanised countries. In 1986 over 85 per cent and 71 per cent of non-aboriginal Australians and Canadians respectively lived in towns and cities (Table 2.4). However aboriginal populations were much more rurally based. More than half of the aboriginal Canadian population lived in small rural communities, including Indian reserves and the small centralised settlements characteristic of the north; and 35 per cent of aboriginal Australians were also rural dwellers. This contrast is particularly important in development terms. Problems stemming from isolation, such as poor access to markets, poor

Table 2.3 Distribution of Indian, Inuit and Metis population, Canada, 1986

Province	North American Indian		Metis		Inuit		Other		Total	
	Nos	% Total	Nos	% Total	Nos	% Total	Nos	% Total	Nos	% Total
Atlantic	26,855	5.1	2,430	1.9	3,995	11.9	1,065	4.4	34,345	4.8
Quebec	62,225	11.8	10,265	8.0	7,050	21.1	1,370	5.7	80,910	11.4
Ontario	146,580	27.9	14,335	11.2	2,270	6.8	4,195	17.5	167,380	23.5
Manitoba	51,470	9.8	28,875	22.5	385	1.2	4,505	18.8	85,235	12.0
Saskatchewan	51,845	9.9	22,300	17.3	100	0.3	3,405	14.2	77,650	10.9
Alberta	62,960	12.0	34,145	26.6	740	2.2	6,080	25.3	103,925	14.6
British Columbia	110,515	21.0	13,065	10.2	740	2.2	2,310	9.6	126,630	17.8
Northern Canada	13,135	2.5	3,130	2.4	18,185	54.3	1,060	4.4	35,510	5.0
Total	525,585	100.0	128,545	100.0	33,465	100.0	23,990	100.0	711,585	100.0

Table 2.4 Australia and Canada, 1986: demographic characteristics

Age group (% total)	Australia		Canada	
	Aborigines	Non-Aborigines	Aboriginal people	Non-Aboriginal people
0–14	39.7	23.1	36.5	21.1
16–64	57.6	66.3	60.1	68.7
65+	2.7	10.7	2.9	10.2
Total	100.0	100.0	100.0	100.0
Crude birth rate	32.0	15.8	25.0	14.0
Infant mortality rate	25.4	9.6	17.0	8.0
Total fertility rate	2.6	1.9	3.0	1.7
Dependency ratio	74.0	51.0	65.0	45.0
Masculinity ratio	98.0	99.0	96.0	98.0
Rural dwellers (% total)	35.0	14.5	53.0	29.0
Urban dwellers (% total)	65.0	85.5	47.0	71.0

communications and high costs for the transport of both people and freight must have a more marked effect on aboriginal peoples than on non-aboriginal peoples. Other less obvious problems concern access to political power and decision-making. People in remote Australia and Canada, whether aboriginal or non-aboriginal, feel marginalised (Young, 1991b). Only a short time in such areas reveals that Canada's north and Australia's north continually indulge in Ottawa and Canberra 'bashing', a sure reflection of people's sense of deprivation, and being passed over by those in control in government. Such experiences are probably even more marked for the aboriginal peoples, isolated not only by distance and the rigours of their physical environment but also by differences in language and in culture, deficiencies in knowledge of how to deal with government or big business and generally made to feel that they belong to the outside.

Demographic characteristics

Aboriginal peoples in both Canada and Australia are considerably younger than non-aboriginal groups, with, in 1986, almost 40 per cent below the age of 15 (Table 2.4). As a result the aboriginal dependency ratios (74 and 65 for Australia and Canada respectively) were considerably higher than those for non-aboriginal populations and in each case over 90 per cent of dependants were children. Such contrasts between the aboriginal and non-aboriginal populations reflect marked differences in fertility. Aboriginal women in

Australia and Canada in 1986 had on average borne between 2.5 and 3.0 children during their child-bearing years, while the equivalent total fertility rates for non-aboriginal groups were 1.9 and 1.7 respectively. Crude birth rates, again much higher for aboriginal peoples, bear out this evidence for higher than average fertility.

Mortality levels also differed. In both Canada and Australia less than 3 per cent of the aboriginal population were aged over 65, while equivalent percentages for the non-aboriginal groups were 10.7 and 10.2 (Table 2.4). Aboriginal infant mortality rates, while lower than in the past, were still relatively high, in the case of Australia almost three times the rate of that for non-aboriginal peoples. Infant mortality rates for Canadian Inuit were also approximately three times the national figure in 1986 (Bone, 1992). As is frequently mentioned in the media, these are unacceptably high figures for rich industrialised countries, well endowed with basic health services which, even for remote dwellers, are almost as accessible as they are in the towns and cities. As many studies now suggest, the persistence of these high mortality rates reflects lifestyle differences, with poor sanitation and lack of healthy diets contributing to aboriginal health deficiencies. It also reflects cultural differences. These often make it very difficult for aboriginal people to use health services effectively (Reid, 1982; Nathan and Liechleitner, 1983a). The introduction of aboriginal controlled health services, such as those now being run by organisations in Alice Springs and in the Pitjantjatjara lands in South Australia, has succeeded in overcoming some of these problems. Studies such as those carried out by D. E. Young (1988) indicate that Canada's high aboriginal infant mortality is due to similar circumstances.

High mortality rates cannot, however, be explained entirely by episodes occurring during infancy. In both Australia and Canada adult mortality rates are also higher than non-aboriginal rates, particularly in younger and middle-age groups. Causes of these differences include diseases resulting from lifestyle changes, such as diabetes; high incidence of alcohol related mortality; and, increasingly, aboriginal adults are succumbing to the same stress related health problems as affect the non-aboriginal population – heart disease and circulatory problems, and cancer. Additional causes of adult deaths include suicide, which for Inuit in 1986 was more than three times the Canadian national rate; and accidents and violence. All of these clearly indicate the stresses with which aboriginal people have to cope in a largely non-aboriginal world. The overall effect of these high mortality rates is low life expectancy. Between 1984 and 1989 Aboriginal Australians had average life expectancies of 55 for men and 63 for women, compared to 73 and 79 for men and women in the national population (Australian Institute of Health and Welfare, 1992). Equivalent figures for Canadian Indians in 1981 were 62 for men and 69 for women, compared to national figures of 72 and 79 (Bone, 1992).

Altogether the demographic profile of aboriginal peoples in both Australia and Canada is more like that of many 'third world' countries than that of the dominant non-aboriginal groups in those regions. As with other socio-economic characteristics, this feature has an important impact on development. For aboriginal people any form of development is obviously affected by the fact that the population is relatively fertile, young, beset by health problems and disproportionately located in more remote and rural areas.

Socio-economic comparisons

Canadian and Australian census data are not entirely appropriate for comparing the socio-economic status of aboriginal Canadians and Australians because the categories and classifications used are not standardised between the two countries. However, accepting such limitations, some socio-economic indicators, such as education, labour force participation, occupational status and income are defined in similar ways and can therefore be easily compared. These indicators are of key importance because they provide clues about peoples' ability to earn sufficient cash to support themselves and their dependants, and also about the formal skills which they can utilise for development purposes. All these indicators reveal similarities between Canadian and Australian aboriginal peoples and at the same time, they provide marked contrasts with the non-aboriginal populations in both countries.

In both Canada and Australia aboriginal young people left school earlier than their non-aboriginal counterparts and consequently far smaller numbers have been able to continue to post-secondary training and gain tertiary or trade qualifications. As Table 2.5 shows, less than 20 per cent of these groups had any recognised qualifications which they might be able to use to join segments of the workforce where higher education is needed; corresponding figures for non-aboriginal groups show that the percentage of those with qualifications was considerably higher, at around 34 per cent. More detailed examination of the levels of qualification, while problematic when comparing Canada and Australia, shows that the proportions of aboriginal peoples who had completed tertiary training (degrees and diplomas) were particularly low

Table 2.5 Australia and Canada, 1986: educational qualifications

(% pop. 15+)	Australia		Canada	
	Aborigines	Non-Aborigines	Aboriginal people	Non-Aboriginal people
With qualifications	10.7	33.6	19.0	31.5
No qualifications	89.3	66.4	81.0	68.5
Total	100.0	100.0	100.0	100.0

Table 2.6 Australia and Canada, 1986: labour force participation and employment

| | Australia | | Canada | |
	Aborigines	Non-Aborigines	Aboriginal people	Non-Aboriginal people
Labour force (% of population 15+)	53.4	60.9	60.3	66.5
Employed (% of labour force)	64.7	90.9	77.3	89.9
Unemployed (% of labour force)	35.3	9.1	22.7	10.1

for Australia (1.1 per cent). The Canadian figure was considerably higher (16.6 per cent), but still much lower than that for Canadian non-aboriginal groups (28 per cent). In contrast, Australian groups had larger percentages with trade qualifications, a result which probably reflects differences in post-secondary educational organisation as well as differences in the data sets. Regardless of these discrepancies, the overall educational disadvantage of the aboriginal groups is quite clear. It has obvious implications in terms of jobs, as well as in terms of potential to grasp whatever opportunity arises.

Aboriginal peoples also generally have lower labour participation rates than non-aboriginal people, although the Canadian aboriginal rate was considerably higher than that in Australia (Table 2.6). On Australian evidence these differences in aboriginal and non-aboriginal labour force participation reflect a number of factors – the problems which aboriginal peoples face in meeting the criteria for registration as an unemployed person and their lack of knowledge on how to apply (Sanders, 1985); and the disproportionate numbers of aboriginal peoples living in remote communities, where job markets are extremely limited and the search for work is known to be futile and hardly worth pursuing. It can be assumed that very similar factors operate for Canadian aboriginal people, particularly in remote areas.

Low labour force participation was coupled with very high rates of unemployment (35.3 per cent and 22.7 per cent for Australian and Canadian aboriginal groups respectively in 1986, Table 2.6). These rates are two to three times greater than the average rates experienced by the non-aboriginal population. The situation in many communities is much worse. Yuendumu (Australia) had an unemployment rate of 66 per cent in 1986 and other communities of similar types would undoubtedly have similar rates. Since then the federal government's Community Development Employment Program (CDEP) has been extended. CDEP participants, despite the fact that their wages are effectively only for part-time work, were not classified as unemployed in 1986. Current rates of aboriginal unemployment, where

CDEP has now been introduced, will therefore probably be considerably lower in many places. Whether this is a cause for optimism or not is a source of some debate. CDEP has been seen by some as a shelving of government responsibility – of making aboriginal people 'work for the dole'. Many Aboriginal communities are however enthusiastic about CDEP. They perceive, quite correctly, that with CDEP they can fund jobs for which otherwise there would be no wages; and that they, and not the government, can decide what jobs are needed. With CDEP, problems associated with unemployment, including alcoholism and general social disruption, have decreased (Sanders, 1988; Morony, 1991).

Low labour force participation and high levels of unemployment, coupled with a lack of post-secondary education inevitably mean that above average proportions of aboriginal peoples are employed in lower-level occupations. In Australia over 34 per cent of aboriginal workers were employed as labourers while the equivalent non-aboriginal group accounted for only 15 per cent of workers. Conversely, while 24 per cent of non-aboriginal workers were classified as managers or professional workers, only 9 per cent of aboriginal people came into this category. The Canadian occupational data make it harder to draw clear comparisons. However it seems that the relative proportions of managers and administrators among aboriginal and non-aboriginal people were 7 per cent and 11 per cent. Lack of people in these categories will obviously affect basic incomes. It also affects people's ability to gain positions of responsibility, and results in the dominance of non-aboriginal people in the top positions even in organisations primarily concerned with aboriginal communities. This applies in both government and non-government situations, particularly in Ottawa, Canberra and main provincial cities. Although departments such as DIAND (Canada) and ATSIC (Australia) follow policies of positive discrimination in employing aboriginal people, most of their sections are headed by non-aboriginal public servants. Even aboriginal organisations, such as land councils in Australia and bodies such as the Assembly of First Nations in Canada employ significant numbers of non-aboriginal people. Although their dominance at top administrative levels is less marked than in the past, they still predominate in positions requiring professional skills (for example lawyers, anthropologists, social science consultants), and, because of the low levels of tertiary education, this is likely to be the case for some time to come.

This situation has profound consequences for development. Regardless of how committed non-aboriginal people are to aiding the cause of aboriginal self-determination, the ideas which flow from them reflect non-aboriginal ways of thinking. All too easily top-down processes of development, which do not take sufficient account of the alternatives suggested by aboriginal people, are accepted and followed.

Limited aboriginal participation in higher-status, well-paid jobs has obvious consequences for people's monetary incomes. As Table 2.7 shows,

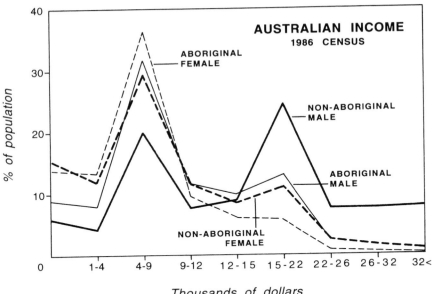

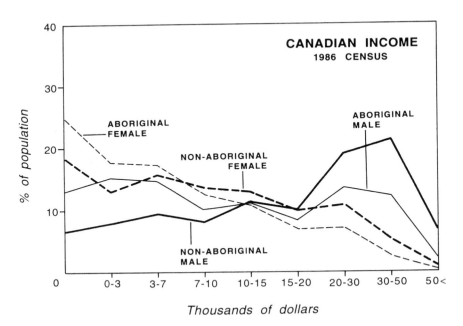

Figure 2.1 Income distributions, Canada and Australia, 1986

Table 2.7 Australia and Canada, 1986: income

	Australia ($A)		Canada ($Can)	
	Aborigines	Non-Aborigines	Aboriginal people	Non-Aboriginal people
$ per capita (pop. 15+)	8,250	12,237	10,740	15,835
$ per capita (total pop.)	4,336	9,018	6,820	12,495

aboriginal incomes in both Australia and Canada were significantly lower than non-aboriginal incomes. In both countries the order of difference is almost identical – the non-aboriginal income per capita for people aged 15 and over in 1986 was approximately 50 per cent greater than the aboriginal income. Figure 2.1 shows that Australia had a rather different pattern of income distribution from Canada. In Australia there was a marked income peak at between $4,000 and $9,000 per annum, a reflection of the remuneration received through welfare pensions and benefits. Not surprisingly aboriginal people, with their high rates of unemployment, figured particularly strongly in this category. Correspondingly, very few of them earned incomes in the upper end of the range. In Canada the peak in the lower-income brackets was less marked, but aboriginal incomes were again strongly concentrated in this part of the range. Few aboriginal Canadians had high cash incomes.

Examining per capita incomes on the basis of the adult population rather than the total population highlights the economic effects of different age structures. For both Australia and Canada the contrast between aboriginal and non-aboriginal incomes increased (Table 2.7). In Australia the non-aboriginal income was now more than double that of the aboriginal income, while in Canada it was close to twice. These changes can be attributed to the high dependency ratios commented upon earlier. Not only did aboriginal people on average receive lower cash incomes but also, because of the youthfulness of the population, those incomes were spread much more widely to support dependants in the family.

On a broad national basis these brief comparisons clearly show the differences in the demographic and socio-economic status of Canada's and Australia's aboriginal and non-aboriginal peoples. Aboriginal peoples are younger, less healthy, less well-educated, poorer and less likely to occupy influential jobs. All of these characteristics foster marginalisation. Their implications for development are obvious.

What of the remote areas in particular? Development indicators, created by combining social and economic characteristics, suggest that their level of disadvantage is well above average. Bone (1992) comments that, in 1981, 80 per cent of the twenty-nine northern Canadian census divisions would be classified as poor or extremely poor, with the Arctic and Sub-Arctic regions containing a very high proportion in the extremely poor category. A study by Khalidi (ATSIC, 1992a) bears this out by showing that all extremely disadvantaged areas were in Australia's Central desert or 'Top End' regions.

Aboriginal access to resources: land

Land, for cultural, economic and political reasons, is the prime resource for aboriginal people and, as summarised above, it is the displacement from the land which has been responsible directly and indirectly for most of the traumas which they face today. For this reason the practical implementation of land rights agreements is fundamental to the whole contemporary development process.

Canada

Land rights agreements in Canada reflect both the change from assimilationist to self-determination development policies, and conflict over the use of resources such as oil, gas, minerals, and water, wildlife conservation and the growth of tourism. In the remote regions these pressures have led to the lodging of claims over much of Northwest Territories, the Yukon and northern parts of Quebec and other provinces (Figure 2.2), resulting, following final agreement, in almost 600,000 sq. km coming exclusively under aboriginal ownership. These claims fall under Canada's Comprehensive Claims policy, based on traditional use and occupancy of the land in areas where aboriginal title has not been previously dealt with by treaty (DIAND, 1981; 1982). The comprehensive policy was applied even in the case of the Dene of the Mackenzie Valley, where Treaties Nos. 8 and 11 were signed in 1899 and 1921 respectively. These treaties, along with any existing title, would be abrogated in favour of new rights legally recognised through the claim. Such new rights are based on protracted negotiations between government and native Canadian groups.

The first Canadian claim to be settled, the James Bay Cree-Naskapi agreements, arose from conflict over the proposed Quebec government development of Cree and Inuit territory for hydro-electric generation. Under these agreements, signed in 1975 and 1978, the Cree and Inuit in the James Bay area accepted financial compensation of $225 million over a 10-year period, outright ownership of 13,300 sq. km of land and hunting rights over a further 135,000 sq. km (Morrison, 1983: 87-95). In return they accepted the extinction of all other aboriginal rights to land. In 1984 the second Canadian claim, the Inuvialuit or COPE (Committee for Original Peoples' Entitlement) claim to the Mackenzie delta and neighbouring areas of the north slope

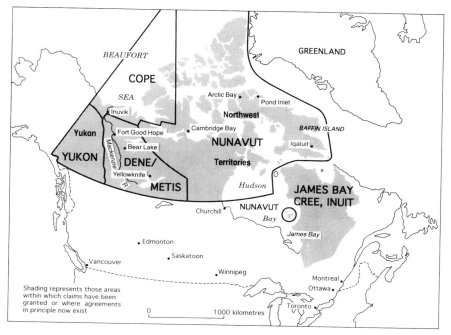

Figure 2.2 Aboriginal land claim areas in Canada

and Arctic islands was finally settled. Its terms included a grant of 90,650 sq. km of land in fee simple absolute, which not only meant that hunting and trapping required native consent but also gave the Inuvialuit Land Administration the right to negotiate rents, participation agreements and other compensation from resource development. Inuvialuit people also gained outright mineral rights to 15 per cent of this area. The accompanying compensation payment, to be paid in varying amounts to the Inuvialuit Regional Corporation over the 15 years from December 1984, amounted to $152 million (DIAND, 1984a: 106).

Both of these claim settlements have been perceived by other Canadian native groups to be deficient, particularly with regard to the extinction of all aboriginal rights, a condition of both settlements. Since December 1986 comprehensive claims have operated under a revised federal policy which enforces surrender of aboriginal title but not of other rights such as freedom in the use of language, or adherence to customary behaviour. Subsequent claims have, somewhat reluctantly, accepted these provisions. Comprehensive claims under discussion also include tacit recognition of the institution of native self-government.

The Dene/Metis claim to their territory in the Mackenzie Valley reached the agreement-in-principle stage in 1988 and, it was expected, would have been completely settled by 1991. That claim contained consistent requests that negotiations include eventual recognition of political autonomy. Their determination stemmed partly from the existence of earlier treaties – Treaty 8 signed in 1899 and Treaty 11 signed in 1921 – interpreted by the Dene as pacts of peace and friendship rather than as a cession of aboriginal rights. Their provisional agreement included the transfer of 181,230 sq. km of land in fee simple, 10,097 sq. km of which would have sub-surface as well as surface rights (DIAND, 1988a). Actual selection of this land was still to be negotiated. Any development of mineral resources on Dene/Metis land to which sub-surface rights were not held by native people would be subject to consultation, as would the plans of the developers themselves and thus it was anticipated that native people would gain considerable control over these activities. Government negotiated royalty agreements within these areas would also be shared by the Dene/Metis, at a rate of at least 10 per cent. This was in addition to a capital transfer of around $500 million within 20 years from the expected settlement date of 31/1/91; $75million of this would be in respect of the Norman Wells proven area, a region of oil and gas development in which some Dene groups were already involved. As with other agreements this settlement included exclusive hunting and trapping rights, and rights of first refusal which would have allowed increased participation in commercial wildlife/fish harvesting and in tourist related ventures associated with renewable resources. This agreement has never been ratified. Negotiations, which were finally broken off in 1990, foundered over the vexed question of the abrogation of aboriginal rights. Subsequently some Dene sub-groups have gone their own way, lodging separate claims (Devine, 1992: 15). This has left the Mackenzie Valley people in limbo, a situation which has come into closer focus since the ratification of the Nunavut claim.

Inuit, like the Dene/Metis people, expressed a determination that nego-tiations over their claim to the eastern Arctic would include the granting of political autonomy. With the signing of the claim agreement in November 1992 they achieved that goal, and the Canadian government is now committed to setting up Nunavut, a self-governing territory which covers the eastern part of the former Northwest Territories. The Inuit claim settlement covers 350,000 sq. km of land, 10 per cent of which is held in fee simple, including sub-surface rights. Mining companies operating within those parts of Nunavut to which Inuit only hold surface rights will have to consult with a surface rights board over issues such as compensation in the form of employment, capital for native business ventures and preferential contracts. Inuit will also expect tourist operators to consult with them over their plans, and discuss possible joint venture arrangements. The Tungavik Federation of Nunavut (TFN), the negotiating body, have also reluctantly accepted surrender of aboriginal title on the grounds that settlement of their claim will legalise their position in those areas granted to them. A capital payment of $580 million over 14 years accompanies the land grant. Additional costs to the Canadian government, for purposes such as setting up a new territorial capital, will be substantial. This is a historic agreement, not only for the Inuit and for Canada but also for other aboriginal peoples seeking political recognition in their own right. Both Jull (1992b) and Fenge (1993) have commented on its significance for aboriginal peoples in Australia, who have become increasingly vocal about their need for political recognition, and who have been watching the Nunavut story with great interest.

The other major claim in process in northern Canada covers the Yukon Territory. The agreement-in-principle, signed in March 1989 and ratified in December 1991, provides for $232 million in cash compensation; 40,000 sq. km of land, 25,000 of which will include sub-surface rights; participation of Yukon natives on various lands and wildlife boards and an obligation on the part of the government to enter into discussions over self-government agreements with any Yukon First nations groups requesting this (DIAND, 1989a).

Although the details of these Canadian agreements vary, there are some important common characteristics. All include specific reference to sub-sistence resources, with extensive harvesting and foraging rights; all include some degree of control over resource development within specified areas, and all include the payment of considerable sums of compensation monies. Theoretically at least these elements provide some basis for development aimed at enhancing economic self-sufficiency.

Australia

As in Canada, the movement for Australian aboriginal land rights was a focal point of the change in policy from assimilation to self-determination, following on from the 1967 referendum. On assuming government after their success in the 1972 federal election the Australian Labor Party (ALP) set the land rights legislation process in motion. However, despite earlier promises to promote such legislation on a national basis this process was restricted to the Northern Territory; elsewhere it has remained a state responsibility. The federal government's failure to exert its powers over the states, seen today by many Aboriginal people as a political sell-out, can be attributed largely to Australia's continuing dependence on primary production and to conflict of interests between resource development, particularly mining, and environ-mental and cultural conservation. This has been particularly important in Western Australia and Queensland where mining and other resource development lobbies are very strong. Federal and state variations in legislation have led to marked inequities between aboriginal groups both in the amount of land which they have been able to claim, and in the types of tenure under which they hold their land. In general people in more remote parts of the Northern Territory, Western Australia and South Australia have

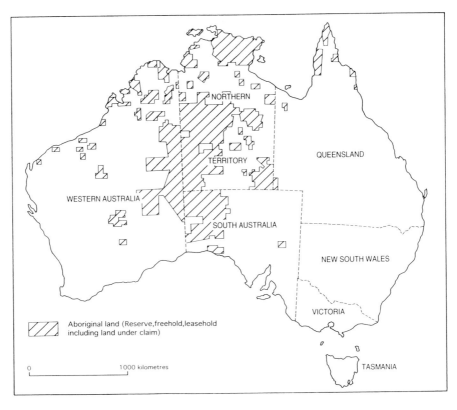

Figure 2.3 Aboriginal lands in Australia

been the most successful in regaining control over large parts of their ancestral territory (Figure 2.3)

All aboriginal people in Australia, given the choice, would prefer to hold the land under freehold tenure, with guarantees over the control of all resource exploitation, including mining. The Northern Territory's land rights legislation, passed in the Aboriginal Land Rights (NT) Act, 1976, provided such tenure and has generally been perceived to be the most favourable to Aboriginal rights and aspirations. Under that legislation all existing Aboriginal reserves in the Northern Territory were immediately granted as freehold land and Aborigines were also able to lay claim, on the basis of traditional affinity to land, to additional areas of unalienated crown land. The Act also allowed for the creation of regional Aboriginal Land Councils to advise on Aboriginal needs and ultimately take responsibility for conducting the claims procedures; and the establishment of a government funded Aboriginal Land Fund Commission to purchase leasehold land to

60

Table 2.8 Aboriginal land in Australia, 1991

	Aboriginal population[a]	As % total population	Total land (sq. km)	Aboriginal freehold area (sq. km)	As % total	Aboriginal leasehold[b] land (sq. km)	As % total	Reserve/ mission land (sq. km)	As % total land	Total ATSI land (sq. km)	ATSI land as % total land
NSW & ACT	70,709	1.18	804,000	492	0.06	842	0.10	–	–	1,349	0.17
Vic.	16,570	0.39	227,600	31	0.01	–	–	–	–	32	0.01
Qld	67,012	2.24	1,727,200	5	0.00	31,990	1.85	95	0.1	33,955	1.97
SA	16,020	1.14	984,000	183,146	18.61	507	0.05	–	–	184,157	18.72
WA	40,002	2.52	2,525,000	35	0.00	103,227	4.09	202,223	8.01	305,485	12.10
NT	38,337	21.87	1,346,200	451,219	33.52	26,424	1.96	0.45	0.00	486,956	36.17
Tas.	8,683	1.92	67,800	2	0.00	–	–	–	–	2	0.00
Australia	257,333	1.53	7,681,800	634,930	8.27	162,990	2.12	202,363	2.63	1,011,934	13.17

Notes: (a) 1991 Census; (b) includes pastoral, special purposes, and local shire council leases.
Source: Heritage Division, Department of Aboriginal Affairs.

meet Aboriginal needs (see below). Land granted under this Act is unalienable freehold, to be held in perpetuity by the appropriate Aboriginal Lands Trusts. Its provisions also confer control over resource development, stipulating that the would-be developers, mostly mineral exploration and exploitation companies, must consult with traditional owners and must draw up a mutually agreed contract before proceeding. This will include the payment of royalties, and provision of employment and training, services and assistance with development. Delays in reaching such agreements have caused bitter conflict between mining companies, the Northern Territory government, Aboriginal Land Councils and the traditional owners concerned. These confrontations have undoubtedly enhanced Aboriginal political experience both at individual and community level but they have also resulted in a 'watering-down' of the original terms of the Act. Nevertheless Northern Territory Aborigines are still able to exert more influence over resource development on their land than are their fellows elsewhere. As a result of the settlement of claims lodged since the late 1970s almost 35 per cent of the Northern Territory is now Aboriginal freehold land (Figure 2.3 and Table 2.8).

Northern Territory land claims are based on proof of Aboriginal traditional ownership to land, expressed both through spiritual knowledge of the area in question and through knowledge and use of the land and its resources for economic subsistence purposes. Thus, as Maddock (1983) and Hiatt (1984) have discussed, the Act incorporates two laws; Aboriginal 'Law', the body of common knowledge and beliefs which expresses not only people's relationship with their ancestral countries but also governs the social relationships and behaviour which form the foundation of their communities; and non-Aboriginal 'Law', in which ownership is expressed through written contract which implies occupational rights and is normally based on exchange of cash for property.

South Australia's land rights agreements, like those in the Northern Territory, also allow for the granting of freehold tenure. However, instead of adopting state-wide legislation, the process has focused on specific claims, two of which have now been granted under the Pitjantjatjara and Maralinga Land Acts, 1981 and 1984 respectively (Toyne and Vachon, 1984). The extensive regions covered by these claims, amounting to over 180,000 sq. km in northern and western parts of the state (Figure 2.3 and Table 2.8) are now freehold land, and traditional owners, like their Northern Territory counterparts, can negotiate with would-be developers for royalty-type payments. However, in contrast to the Northern Territory, there is no provision in South Australia for Aborigines to make further claims for additional areas of vacant crown land. Their current holdings could only be extended through new legislation. Many South Australian Aborigines thus remain landless.

Freehold title is also granted under New South Wales legislation and existing reserves were all converted to this form of tenure. However, because of their small size, this has resulted in very restricted amounts of Aboriginal land throughout the state (Table 2.8) and although the Act has allowed for additional claims to areas of unalienated crown land, relatively few of these have yet been successfully negotiated.

Other forms of aboriginal land tenure include leasehold and reserve, mostly within Western Australia, Queensland and the Northern Territory (Table 2.8). Aboriginal land in Queensland, which presently amounts to only 2 per cent of the state, consists largely of community reserve lands held as perpetual leasehold under Deeds of Grant in Trust (DOGIT) to elected councils of Aborigines. This form of tenure is, in aboriginal eyes, deficient because it still allows for government interference, especially to facilitate resource development. With the 1989 change in state government from a conservative coalition to the ALP, aboriginal people hoped for radical reform. However they have been very disappointed with the new legislation which, by extending the title of land which they already hold under reserves, will only give them control over a further 1 per cent of the state. While Aboriginal communities holding that land will be able to recommend against mining, forestry or other resource exploitation, their desires could be overruled by the Queensland State Cabinet. This restrictive situation reflects Queensland government concern over the possible impact of Aboriginal land claims on its vital mining and pastoral industries; its perceived political vulnerability; and its problem with determining how claims could be proven by people who were in many cases moved from their original lands many years ago. Not surprisingly Aboriginal disappointment with this legislation has been strongly voiced, and potentially violent public confrontations have already occurred.

Aboriginal land in Western Australia is held under reserve and leasehold titles (Table 2.8). In that state pressures exerted by the powerful mining lobby were a vital factor in blocking the passage of legislation. Western Australian reserve lands have since remained under the control of the state government and in theory could be externally administered with little Aboriginal input into resource use decisions. These lands have now been vested in a government appointed Aboriginal Lands Trust, which in turn may lease land to communities for ninety-nine years. As Aborigines have made abundantly clear (for example, Coombs et al., 1989), they still feel that this type of lease provides insufficient guarantee for the future, and that the ever present threat of unsolicited mineral exploitation on Aboriginal lands, highlighted in the Noonkanbah confrontations of the 1980s (Hawke and Gallagher, 1989) still hangs over them.

Neither Victoria nor Tasmania have yet passed land rights legislation, although Aborigines in both states have clearly expressed their need for land, and for Land Councils to administer both the land and the communities on it. Thus Victoria's grant of title over the former reserves of Lake Tyers and Framlingham is at present as far as the process has gone in that state.

Tasmania's Land Rights Bill was defeated in Legislative Council in July 1991 and Tasmanian Aborigines have yet to receive title to areas such as Oyster Cove and Bass Straits islands both of which are of past and present significance to them.

Leasehold land also includes pastoral leases purchased for aboriginal groups largely under the federal government's land purchase programs. Leasehold rights are generally more restrictive. In the Northern Territory most Aborginal-held pastoral leases have already been converted to freehold under the terms of the Act; but in other states such as Western Australia, where no such provisions for conversion exist, severe restrictions may apply. There Aboriginal pastoralists are forced to comply with rules governing stocking levels and the maintenance of improvements such as fences, bores and wells and are indirectly hindered from introducing alternative forms of land use. Moreover they do not have the right to deny access to outsiders with entry permits, mining companies can hold exploration and development licences within their lease boundaries, and developers are not obliged to negotiate royalty agreements or pay compensation.

In June 1992, after ten years of waiting, the High Court of Australia found in favour of a claim by Eddie Mabo, a Murray Islander from the Torres Straits, that he and his fellow islanders had rights of prior occupation to that land, and that those rights had not been extinguished by subsequent alienation. This historic decision, which effectively recognises native title for aboriginal Australians and at last negates the concept of *terra nullius*, has been greeted with much enthusiasm and has lent new momentum to the land rights movement. At last it effectively brings Australian recognition of the rights of its 'first peoples' into line with that of other former British colonies – USA, New Zealand and Canada.

After wide-ranging discussion and consultation with aboriginal groups, state and territory governments and other key organisations such as the Australian Mining Industry Council, the National Farmers' Federation and the Australian Conservation Foundation the federal government responded in 1993 by presenting draft legislation for a Native Title Bill. The passage of this legislation, incorporating amendments to deal with both Aboriginal and non-Aboriginal concerns about land tenure and management, in December 1993 was greeted with much relief and hope for the future. The year 1994 has seen the establishment of tribunals to deal with land claims under native title. Since these claims must be based on continuing traditional knowledge and contact with the land and will apply mainly in areas of unalienated crown land the benefits will not be evenly spread. Western Australia, which has huge areas of such land and which currently lacks legislation conferring control over resource development is likely to be a key area. As Peter Yu, Director of the Kimberley Land Council, commented at a recent forum in Darwin, 'Mabo is so important for us because it returns the argument for self-determination to its real meaning – that is about the right of a dispossessed people to their territory' (Yu, 1994). Present outcomes here are uncertain because of the hurried passage of Western Australian legislation to extinguish native title throughout the state. The federal government's assurance that their legislation will override this remains to be tested. Those who cannot gain much from the Native Title legislation will not necessarily be forgotten. The federal government's 1994 budget, brought down in early May, has also addressed Aboriginal social justice issues, including an Aboriginal National Land Fund of $1.46 billion over ten years. Those whose claims will be hard to justify under the Native Title legislation will be the major beneficiaries from this package.

Aboriginal people and resources in remote Canada and Australia

The general picture which emerges from the preceding discussion is that the indigenous minority peoples of both Canada and Australia are clearly disadvantaged in socio-economic terms compared to the rest of the population. This generalisation tends to give an impression of uniformity which disguises the socio-economic diversity which exists within the aboriginal population groups. Diversity is also a feature of their degree of control over land and resources, shown both through their opportunities to make claim to land and the differing provincial, state and territory legislations which control the types of claim which they can successfully make. This study is specifically concerned with those aboriginal groups living in remote parts of these countries, areas where much of the land is environmentally marginal and has limited natural resources. Here the essential relationships between the people and their environment, and the diversity of these relationships is more effectively brought out through descriptions of their settlements.

Aboriginal settlements

The rural concentration of population is particularly marked in Canada's north and in Australia's centre and north. In 1986 over 80 per cent of the aboriginal population of Northwest Territories and Yukon lived outside the main towns such as Yellowknife and Whitehorse; the non-aboriginal equivalent was only 28 per cent. The Australian Northern Territory's Aboriginal population in that same year was also strongly rurally based, with 70 per cent outside the towns. Differences, both between town and rural settlements, and between types of rural settlement also influence development.

In both Canada and Australia the current aboriginal settlement pattern in remote areas is largely a product of the history of contact with non-aboriginal settlers. It has been very strongly influenced by the growth of towns, both those concerned with service delivery and those, like mining towns, which are the product of resource development. As elsewhere, these towns have attracted aboriginal people, often younger people interested in breaking away from some of the restrictions of traditional society and wanting to involve themselves in the 'whiteman's world'. Assimilation policies, implemented through the deliberate gathering together of small groups of aboriginal people into centralised settlements where they could be housed, educated and cared for, have also had a significant effect. And, in Australia, where the pastoral industry relied heavily on an Aboriginal labour force, Aboriginal communities became established on many of the properties. Overall the earlier aboriginal pattern of land occupation, which was much more fluid and was characterised by small extended family groups, was replaced by a strongly centralised settlement pattern which reflected conventional non-aboriginal sedentary forms of existence.

Since the 1970s land rights, and the resultant reoccupation of the land, has changed this established pattern, particularly in Australia. Here, as many have documented (for example, House of Representatives Standing Committee on Aboriginal Affairs, 1987; Nathan and Liechleitner, 1983b; Young and Doohan, 1989; Taylor, 1992), Aboriginal people have moved back on to the land, establishing 'outstations' or 'homeland centres' back in country which, for some of them, had rarely been visited in decades. This movement, which is clearly at odds with that common to other parts of the country where the urban centres, both large and small, continue to draw people from the bush, is a reaffirmation of their social, cultural and economic ties to the land. Remote Canada has no exact equivalent. This must partly be due to the environment, which precludes easy movement for large periods of the year. However here many of those living in the centralised northern settlements also spend significant periods of time out in smaller settlements, outpost camps in the boreal forest or the tundra where, from their traditional country, they can involve themselves more fully in hunting, fishing and trapping. As in Australia, these small scattered communities maintain strong and vital links with the centralised settlements and the northern towns and movement of people between them is constant. Although all these types of remote aboriginal community are linked by the social networks and constant mobility of the people, they still have different characteristics. These exert a strong influence on people's lifestyles, their education, their experiences of wage employment and their social coherence, and hence affect their response to development opportunities. The following more detailed descriptions, relating specifically to the regions within which I have conducted fieldwork and many of which are subsequently referred to in the text, bring out these contrasts more clearly.

Canada

Yellowknife, with less than 10,000 people in 1986, is by far the largest town in Canada's Northwest Territories. None of the other regional centres, such

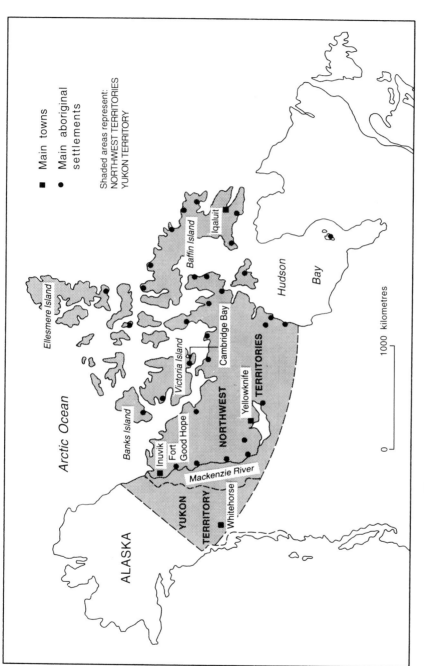

Figure 2.4 Remote Canada: main settlements referred to in text

Plate 2.1 Yellowknife, administrative centre of Canada's Northwest Territories

as Frobisher Bay (now Iqaluit) and Inuvik, exceeded 3,500 people. Outside these towns, however, a centralised settlement pattern prevailed, with over forty such communities scattered through the region (Figure 2.4). Some of these small 'towns' had total populations exceeding 1,000. Many of the earliest of these towns grew from trading posts run by the Hudson's Bay Company, while some of those established in the twentieth century were set up under government auspices to assist in the process of assimilating aboriginal people to non-aboriginal lifestyles. Fort Good Hope and Cambridge Bay provide good examples of these two different types of community.

Fort Good Hope

Situated on the lower Mackenzie River (Figure 2.4), this started as a trading post with the North West Company in the early nineteenth century and subsequently came under the Hudson's Bay Company. In 1986, 92 per cent of its population of 550 were aboriginal, Hare-speaking Dene Indians originally from the surrounding boreal forest regions along the river. Hunting and trapping had dominated the lives of these families for generations, and many strongly retained their interests in the land and its resources. At the same time most alive today were born and bred in Fort Good Hope, and have become accustomed to living in well appointed log houses with modern services. They have gone to school, both within the village and also, in more recent times, to high schools elsewhere, mostly

Plate 2.2 First snow at Fort Good Hope. Note substantial log cabins, with tipi smokehouse (for curing skins) in the yard

Plate 2.3 Village transport in the sub-Arctic and Arctic – four-wheel cycles, used for carrying people and freight across the tundra

boarding establishments run by the Catholic Church. Fort Good Hope also had a Hudson's Bay Company store, a health clinic, handcraft centre and the Hunters and Trappers Association (HTA). These provided a few permanent jobs, and employment opportunities were occasionally augmented by casual work for local mining and exploration teams. The Band Council was responsible for the town's administration and also development affecting Fort Good Hope people. Recent development included the negotiation of joint venture agreements for oil and gas exploitation (see pp. 173–4) and plans to establish a hotel, so that the community could begin to attract tourists.

Although Fort Good Hope people saw themselves primarily as town residents they still retained their interests in living off the land. Most of them only left town in the summer months, when they moved out to fish camps to take advantage of the rich resources of the Mackenzie River during its short open period (see pp. 133–4). Some, however, were more strongly committed to the outdoor life. They, or at least some members of their families, worked as hunters and trappers both in summer and winter, an occupation which required them to spend lengthy periods in remote camps in the surrounding forest. This outcamp movement, which involved about a quarter of Fort Good Hope's population in the winter of 1984, not only demonstrated people's determination to continue using their traditional skills and look after the country, but also helped them to reassert their identity and desire for independence. It was also seen as a way of avoiding some of the social problems common in the larger community. Outcamps were discouraged under assimilationist policies. However during the 1980s their social and economic importance had increasingly been recognised (GWNT, 1981) and the local Hunters and Trappers Association (HTA) had received some funding from GNWT and the Band Council to assist in moving their families to their traplines. This was costly – hire of a float plane to ferry people and equipment out before the winter freeze was complete; and advance purchase of equipment and food. Hunting families also had to raise their own funds to cover expenses, through bingo nights and through their own savings. Because of their commitment to that way of life, which they clearly felt was justified, the centralised settlement pattern of the Fort Good Hope area had begun to break down into one which combined permanent and more temporary settlement. This is a compromise between aboriginal and non-aboriginal forms of land occupation in the northern boreal forest.

Cambridge Bay

On the southern shores of Victoria Island in the Canadian Arctic, this is a much more recent settlement than Fort Good Hope. One of the largest Inuit communities in Northwest Territories, its 1986 population was around 1,000, 77 per cent of whom were aboriginal. As a relatively new town in the 1950s

Plate 2.4 Cambridge Bay house, triple insulated and with small windows to prevent excessive heat loss. Note fish drying rack on exterior

and 1960s, Cambridge Bay owed its origins partly to the establishment of a station on the Distant Early Warning (DEW) line system which stretched across northern Canada. Since the 1970s and particularly in the 1980s its dominant role has been that of a service centre, and the seat of regional administration for the Kitikmeot area of NWT. This reflects GNWT's policy of decentralising its offices, as a deliberate attempt to improve services to its extremely isolated and dispersed population. Cambridge Bay's large non-aboriginal population and the role of GNWT as an employer (providing jobs for one-third of all the town's workers) provided obvious contrasts with Fort Good Hope.

Unlike the older Dene, many of the older Inuit at Cambridge Bay had grown up on the land rather than in settlements. Assimilation and increasing dependence on welfare forced them to move to the town. Subsequently they had become accustomed to the more sedentary lifestyle, and wanted housing, schooling for the children, health services and jobs. Because of its regional role Cambridge Bay's biggest employer was GNWT. Other service establishments, such as the Hudson's Bay Company store and the local Inuit co-operative provided most other jobs. The co-op, in addition to a retail store, ran a wide range of other businesses including an arts and crafts centre, a fish freezing plant, a bakery, a hotel and a taxi company. With up to sixty workers during the fishing season it was the biggest Inuit employer in town and one of the leading co-ops in the whole Nunavut region. Other Inuit organisations in Cambridge Bay included the Kikitmeot Inuit Association (KIA), a local political group concerned with the Nunavut land claim negotiations; and the recently formed Enokhok Development Corporation (EDC), an organisation promoting Inuit economic self-sufficiency. Eno-khok's proposals included the development of office buildings for GWNT and others; the establishment of small businesses such as a hairdresser and a stationery store etc. EDC expected to apply for government funding, but also planned to collaborate with TFN to ensure that they would be part of the economic development effort which would follow the completion of the Nunavut claim. Enokhok's role was seen as complementary to that of the GNWT Department of Economic Development, also a major force for the promotion of self-sufficiency in Cambridge Bay but, in 1984 at least, more strongly focused on non-aboriginal small business development.

Although predominantly townspeople Cambridge Bay's Inuit residents also moved out to live partly off the land in outcamps. Summer camps were primarily for fishing and in winter, when overland travel was much easier than in summer, people also obtained furs and pelts. Although these campers could be relatively self-sufficent they also maintained regular contact with Cambridge Bay and, as in Fort Good Hope, they received some assistance from GNWT. Outcamps were very important to people and the 1980s collapse of the sealskin markets had been a cause of great concern.

Australia

Aborigines in remote Australia live in five main types of community: Aboriginal towns, the former government and mission settlements; on Aboriginal-owned cattle stations; on non-Aboriginal cattle stations; in homeland communities, or outstations, mostly located on Aboriginal-held land and largely made up of people closely related to these specific areas of country; and in established non-Aboriginal towns. Yuendumu, its associated outstations, neighbouring cattle station communities and Alice Springs provide a composite example of all types (Figures 2.5a and 2.5b).

Yuendumu

About 300 km to the north west of Alice Springs, Yuendumu was established as a government settlement in the 1940s to bring together the Warlpiri, Pintupi and Anmatyerre people from the Tanami region and assimilate them with non-aboriginal society. Housing, schooling, attendance at health clinics, learning how to use western foods and shop in western stores, and replacing Aboriginal spiritual beliefs with the doctrine of the Baptist church were all part of this system. By the early 1970s Yuendumu had become a small town

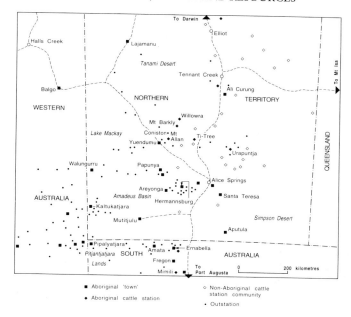

Figure 2.5a Central Australia: distribution of Aboriginal settlements

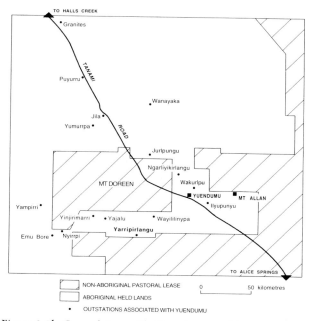

Figure 2.5b Central Australia: Yuendumu and its outstations

of around 1,000 people, approximately 85 per cent of whom were Aboriginal and many of whom had, to the outside observer at least, come to favour the non-aboriginal way of life. Many of Yuendumu's residents had never lived anywhere else.

The 1970s and particularly land rights wrought great changes. Through the NT Land Rights Act Yuendumu families regained title to much of their traditional country – the reserve around the town, other huge reserves to the west and northwest and, under claim as unalienated crown land, the Tanami Desert itself. As their land has been returned, people have begun to move back, setting up their first outstations in the mid-1970s; by the late 1980s between 15 and 20 of these remote communities were firmly established. Because of this movement the population of Yuendumu itself declined to around 600. All the basic services were still there – a school, now taking not only children but adults up to and including tertiary level; a clinic; a housing company, responsible for managing a stock of over 100 dwellings; several stores, one of which was a large supermarket; an arts and crafts centre; a mining company; and a cattle company. Most services were provided by the government; however all companies and businesses were Aboriginal owned, albeit mostly with a high level of government funding. Yuendumu Community Council, a representative elected body, was responsible for central administration. The town still had its quota of non-Aboriginal residents, some long-term and some recent incomers and mostly holding key positions in services or businesses. However Aboriginal takeover in some of these roles was beginning to increase, and the levels of formal education were rising.

The town has always faced both social and economic problems. These can be partly attributed to the fragmented nature of the population, three major language groups with a number of sub-groups of people from different areas. The lack of an economic base, and hence high unemployment, has been another factor. Not surprisingly, conflicts, including fighting and alcohol abuse, have all been common. Those instabilities, along with the return of the land and the people's increasing determination to take control of their own destinies, underlie the development of Yuendumu's outstations. Other reasons for outstation growth have included the limited value of subsistence resources around the town itself and people's desire to reforge their cultural links with country.

Yuendumu's outstations

Yuendumu's outstation resource centre now co-ordinates the service needs of about a dozen small settlements lying to the west and north of the town. Most lie on or near the Tanami Desert and all are on Aboriginal freehold land. Their residents have chosen locations as close to their ancestral countries as possible, within the limits imposed by water supplies. Initially these outstation groups had a very basic lifestyle – camping in the open

Plate 2.5 Lifeblood for a Tanami desert outstation: permanent water, with bore capped with hand-pump

behind protective windbreaks, transporting water from town or from nearby springs, very limited communications and no other services. Government support was based on the principle that they had to prove that they were determined to stay on the land before any funding would be provided. Nowadays, as their populations have grown, they have not only acquired bores and tanks but also housing, small clinics and stores and, in some cases, bush schools. Many now have STD telephones. The largest of them, Nyirrpi, has grown so much that it is now funded by government as a community in its own right, an Aboriginal town of about 250 people. It also now boasts satellite outstations.

In contrast to the Canadian outcamps, these outstations are not seasonal settlements but are occupied all the year round. Although the population is highly mobile each has a core group who identify strongly with the surrounding country, and subsistence and associated ceremonial activity is very important to them. Outstation infrastructures are no longer those of a conventional nomadic population. Wage employment prospects remain very small but the introduction of the Community Development Employment Program (CDEP), a special federal government program whereby the community is funded on the basis of unemployment benefit equivalent for the provision of part-time jobs (see pp. 116–17), has helped to provide service workers. These jobs can include driving the community vehicles, maintaining the water supplies or running the radio schedules. In many outstations the art and craft industry, conducted with the help of the Yuendumu-based

Plate 2.6 Wakurlpu outstation, near Yuendumu

Plate 2.7 Outstation school

Warlukurlangu Artists Association, is the main industry.

Although the outstation movement appears to have been socially beneficial to many Aborigines, it has presented problems to government service agencies. They see the costs of supporting people in remote country as prohibitive and as a result many outstations still lack schools, health services and basic essential services. They may also lack communications, and the need for robust community vehicles has been a major emphasis of such desert groups from their very beginning. There are also, in such remote and harsh areas, some physical dangers. As recently as the summer of 1991, following a vehicle breakdown on a road west of the Tanami, eight Warlpiri died of dehydration.

Mt Allan

An Aboriginal owned cattle station with about 2,360 sq. km of land, Mt Allan lies directly to the east of Yuendumu. When it was purchased by the federal government's Aboriginal Land Fund Commission in 1976 it had a population of around sixty Anmatyerre, the traditional owners of the land. Since 1946, when Mt Allan was established, they had formed the core of the stock-camp and their local knowledge both of the environmental and cultural characteristics of the country had played an important role in the success of the enterprise. Following the 1976 purchase other Anmatyerre families who

Plate 2.8 Aboriginal camp at Mt Allan

had worked on surrounding stations returned to their country and by the early 1980s the population had doubled in size. With the assistance of a non-Aboriginal manager, the people continued to run the station as a commercial enterprise and efficiently upheld all the requirements of their lease. During the 1980s Mt Allan people went through a land claim which was based on their traditional ownership of the country (Young, 1987a), and in 1988 they finally received freehold title to the land. Subsequently there have been some settlement readjustments. Some families, for reasons similar to those already highlighted for Yuendumu, have left the central homestead to establish their own outstations within the former lease. This assertion of cultural identity has not only led to population dispersal but has also affected commercial aspects of the Mt Allan enterprise. People have shown an increasing interest in alternative forms of land use, including subsistence, and have also developed their own locally based arts and crafts industry. Altogether the whole community has been undergoing a readjustment to the realities of land ownership.

Napperby station

To the east of Mt Allan and in non-Aboriginal ownership since the early twentieth century, Napperby station has an Aboriginal community of about 200 people, Laramba. Like their kin on Mt Allen they are traditional owners of the country within the Napperby lease and since the early days of the station they have worked with the cattle. However, until recently they have had no secure tenure to the land where they lived. Their application for a small area of Napperby to be excised as a living area for them, a protracted affair because of strong opposition from the non-Aboriginal pastoralist who saw it as a threat to the commercial viability of the whole station, has only recently been granted. While the area they have received was not their first preference (their prime interest was in the site of the homestead itself, a permanent waterhole of great cultural significance), the granting of the excision has enabled them to get better housing and some services; however it has not solved all their problems. There are few jobs; and their access to subsistence resources is sometimes restricted because, according to the non-Aboriginal pastoralist, this might interfere with the operation of the cattle property. Altogether places like Laramba are generally among the most disadvantaged in the Aboriginal community. In recent times, however, the Laramba mob have also begun to reap some economic benefits from arts and crafts production.

Alice Springs

Alice Springs has about 5,000 Aboriginal residents, approximately one-quarter of the town's population. More than half of these families have been town residents for generations, and have fitted in with the rest of the population, living in conventional private or public dwellings and being part of the workforce. However many others moved into Alice Springs in the late 1960s when the granting of Award wages in the pastoral industry led to their eviction from surrounding cattle stations. Over the next decade many of these families camped out in the dry creek beds, with no housing, no facilities and under the continual threat of being moved on by the police. They had nowhere else to go. Today, thanks to the efforts of their own self-help organisation, Tangentyere Council, their conditions are much improved. They live in approximately twenty-five town camps, each on a separately negotiated lease on which they have constructed their own community; most have satisfactory housing and services; and the town campers as a group have established their own primary school. Health care comes from another Aboriginal organisation, and the commercial TV licence for the whole of Central Australia is held by yet another Aboriginal company, a media association. Together these organisations provide the bulk of Aboriginal jobs in the town.

Towns like Alice Springs are cultural melting pots, both from Aboriginal and non-Aboriginal viewpoints, bringing together different language groups and people who otherwise would have had little contact with each other. Social unrest and problems arising from the readjustment of these different groups all occur, causing not only minor conflict but excessive drinking and severe violence. Racism is also an all-pervading attitude. These things are hard for the Aboriginal residents of Alice Springs to cope with. They also cause problems for Aboriginal people from outside, the people of places like Yuendumu and Mt Allan who are transient visitors, coming to town to see their relatives, attend meetings or the hospital, deal with business, or just to have a spree.

Categorising contemporary Aboriginal communities in this way helps to focus on characteristics which affect development. Although the relative importance of these has rarely been measured, Aboriginal communities can be thought of as a spectrum ranging from outstations, the most traditional and least involved in the cash economy, to the Alice Springs residents, many of whom are well integrated into Australian urban life. Fisk's (1985) estimates of relative incomes suggest that, while outstation groups were the poorest in terms of monetary income, town dwelling groups who have very limited access to subsistence resources are overall the most economically disadvantaged when all income sources are considered. For all of these groups the retention of Aboriginal identity is of prime importance.

Aboriginal lands and development

In both Canada and Australia the largest areas of aboriginal land lie in the more remote regions. Aboriginal peoples who live in southern Canada or

Australia have gained very little and urban aboriginal people as a whole have missed out in this important process. People in the Canadian north or the Australian outback have, in contrast, been fortunate and in relation to their populations have now regained control over significant parts of their territories. However that has occurred largely because their homelands were perceived by the non-aboriginal population to be of least value for commercial development. In Australia over three-quarters of aboriginal held land falls into arid and semi-arid zones, with less than 600 mm of annual precipitation. Much of the remainder lies in the tropical north, where the unpredictable monsoon climate also frequently results in pronounced dry seasons and leads to environmental problems very similar to those of the arid interior. Thus, for most of these lands, poverty of soils and vegetation limit their economic development potential. This is compounded by a high degree of physical remoteness. Mineral and other natural resources, while not yet fully documented and certainly more extensive than earlier anticipated, also do not appear to be extensive. Canadian aboriginal lands are also en-vironmentally disadvantaged – harsh northern lands, with short cool summers and long bitterly cold winters, beyond the limits of commercial agriculture and large-scale forestry and extremely inaccessible except by expensive air travel. Their known resources are largely restricted to minerals and to wildlife in terrestrial and marine environments. Thus, in both Canada and Australia, the relative advantage which remote dwelling aboriginal people hold in terms of land is tempered by the limited value of that land as a development resource.

The economic effects of land claim agreements

In both Australia and Canada the settlement of aboriginal land claims has had important economic implications for those involved. These include issues relating to control and use of renewable resources; those concerning the granting of sub-surface and surface rights; those covering the payment of cash sums in compensation for land lost and the extinguishment of earlier rights; and the rules governing the use and distribution of these cash payments. All agreements recognise aboriginal rights to hunt and forage freely on land granted to them, although such rights may be more restricted on federal held lands; and many, such as the more recent Canadian agreements, give preference to aboriginal land-owners in the development of natural resources. Protection of subsistence, and hence of its economic contribution, is thus apparently assured. Similarly, Canadian and Australian agreements include some areas of sub-surface as well as surface rights, thus forcing developers to negotiate with aboriginal owners before they can proceed, and ensuring that, where relevant, royalty-type arrangements will be made. In both countries granting of sub-surface rights to areas of proven mineral wealth has been excluded from claims, but the possibilities of subsequent discovery of deposits in areas selected is high. These provisions should help to promote economic self-sufficiency outside the government welfare sector. Methods used include the use of royalty monies to finance new projects, purchase shares in existing commercial operations or even establish investment corporations to support future development. In Canada, where the comprehensive land claim agreements have included substantial compensation monies, all three options are possible. In Australia's pre-Mabo era, where no compensation monies have been paid, investment corporations have not been a possiblility.

Two Canadian claims, the James Bay and Northern Quebec agreements and the Inuvialuit (COPE) agreement, have resulted in the establishment of aboriginal investment corporations. The Makivik (for the Inuit) and James Bay Regional Authority Board (JBRABC) (for the Cree) corporations, which resulted from the James Bay agreements, are non-profit-making bodies involved in the investment of compensation payments and in financing their own community development. As Robinson *et al.* (1989a) point out, these aims can conflict with one another, because some claimants may well prefer the money to be spent on community development while others are concerned with investment for the future. The Inuvialuit Development Corporation (IDC) was incorporated in 1985 following settlement of the COPE claim. Up to 1987 it had made investments in retailing and wholesaling; oil and gas exploration and development; real estate; and transportation. Since these investments are wide-ranging and also locally based, Robinson *et al.* (1989a) felt that IDC had created a more secure financial base than would have been possible if they had limited their interests. The experiences of these two organisations, along with the much longer established Alaska Native Corporations (Pretes, 1988b; Wuttunee, 1988; Pretes and Robinson, 1989c; Robinson *et al.*, 1989b), provide important prototypes of what may eventually become the structure in Nunavut. If Australian Aboriginal groups are able to use the Mabo case to make successful compensation claims in future, they will also assist the development of similar investment organisations 'down under'.

The other main economic benefit from land claims comes from royalty monies. These can provide substantial sums which, depending on the type of agreement reached by those concerned, can be distributed to individual claimants, can be used for general community benefit, or can be invested. In Australia the Gagudju and Gunwinku Associations, recipients of royalty monies from uranium mining in the Alligator Rivers region in the Northern Territory, provide the most detailed examples of the benefits and costs arising from the use of royalty monies (see O'Faircheallaigh, 1988; and Chapter 5, below). In general terms the Northern Territory's Aboriginal Benefit Trust Account (ABTA), into which the uranium royalties are paid, has provided substantial assistance to a wide range of communities (Altman, 1985). Its responsibilities cover not only the traditional owners in the areas where mining is actually occurring, but also support of the Aboriginal Land Councils and the administration of a fund to be used for the benefit of all Northern Territory Aboriginal groups. By 1990 ABTA had received over $200 million, principally from uranium mining royalties, over $47 million of which had been disbursed as grants to groups all over the Northern Territory.

Other financial resources available to aboriginal groups in remote areas

Economic development could, theoretically, also be supported through the private sector, using conventional financial organisations such as banks and investment companies. However this is not an option for many aboriginal groups, particularly those in remote areas. Apart from those who have benefited through land claims, their financial prospects are extremely poor. Not only do they suffer the common problems of lack of financial equity and knowhow, but they are physically remote from the system. As a brief study by Stanley (1982) described, banks are generally loath to locate themselves in remote Aboriginal communities because the nature of financial business common to such groups – small and frequent transactions with little long-term investment – is not profitable for them. Centralisation of banks in towns, mostly far distant from such settlements, coupled with problems of communication inhibit access. While collaboration between public and private funding groups has occurred (see below, Chapter 3) this has been rare, and it has been more a case of using the regional expertise of the banks.

While it is currently unlikely that people in remote communities will be successful in raising capital through the private sector, this may change in the future. Examples of privately established aboriginal financial institutions have already appeared in urban communities, such as the Mohawk community of Kahnawake, on the outskirts of Montreal. As Ponting (1986: 166–7) discusses, the Mohawk Council in 1984 decided that they should create a 'people's bank', affiliated with the Caisse Populaires Desjardins, as a local institution controlling the circulation of money within the community. In that year $7 million in cheques had to be cashed for Kahnawake people, all of it through organisations outside the community. This potentially filtered off much of the local benefit. By 1989 the Caisse Populaire had been in existence for five years, but fully operational for only about three. It had received initial support in the form of an operational loan and insurance from the Quebec federation of Caisse Populaires and was operating successfully. Most of its business concerned normal banking activities, such as cashing cheques, depositing, etc. but twelve business loans had been granted and it was anticipated that this side of the business would expand (Stephen Horn, pers. comm., 1989). A 'people's bank' of this type could well play an important role in the distribution of funds coming through compensation land claim payments in remote Australia and Canada in the future.

MODELS FOR ABORIGINAL DEVELOPMENT IN REMOTE AREAS

Aboriginal people in remote parts of Canada and Australia are hindered from effective participation in development for a number of reasons: their physical remoteness and the poverty and geographical constraints of the environment in which they live; cultural traits which clearly differentiate them from fellow non-aboriginal remote-dwellers and, even more clearly, from the urban populations which dominate both of these nations; low socio-economic status and an entrenched dependency which is the legacy of many decades of suppression and dispossession; and, although land rights movements have increased their control over resources, the value of these resources in economic terms is mostly either small or unknown. Conventional models of development, particularly those which focus primarily on economic growth and large-scale resource exploitation, provide little understanding of the present or guidance for the future. Most of the development models which have so far been discussed take alternative approaches. They include the model of the 'informal economy' and that of the 'next economy'.

The first model, that of the 'informal' economy, highlights components of the aboriginal economy commonly ignored in the implementation of government policies and opens avenues for finding ways of incorporating these. For well over two decades anthropologists, geographers and other social scientists (for example, Usher (1980); Dacks (1981); Feit (1982b)) have commented about the importance of non-monetary components in the aboriginal economy in Canada's north, and, by implication, on the need to take this into account in northern development policy. Similar comments for Australia have come from Coombs *et al.* (1989); Altman and Taylor (1987); and Young *et al.* (1991). Usher describes this as the informal economy of the north. As he rightly stresses, it can provide a vitally important component of sustenance and survival for aboriginal people. Not only do Indian and Inuit harvest country foods for their own consumption, thereby, as Quigley and McBride (1987), Usher (1982) and others have shown, accounting for significant proportions of their food intake, but the hunting and trapping industry has for generations provided their main source of cash. In northern Australia some favoured Aboriginal groups living in coastal or richly endowed inland riverine regions can, at least during some seasons in the year, provide over 80 per cent of their protein requirements from hunting, fishing and foraging (Meehan, 1982; Altman, 1987). Even in the less well endowed central desert Aborigines supplement their purchased food supplies with significant quantities of game, fruits and vegetables (Cane and Stanley, 1985; Devitt, 1988). Bush materials in both areas also provide the wherewithal to make clothing, tools and other equipment, and through industries such as arts and crafts, allow the linkage into the cash economy. Thus in both the Australian and Canadian cases the informal economy is interwoven with the

83

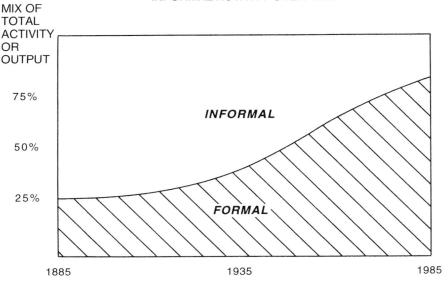

Figure 2.6 The changing mix of formal and informal activity over time
Source: Adapted from Ross and Usher, 1986: 53

formal, or cash economy, a characteristic the importance of which Ross and Usher (1986) rightly stress. Their discussion of formal and informal economies refutes the concept of dualism, whereby these processes remain distinct from one another, and instead presents the concept of interlinkage, akin to Brookfield's interdependent development (1975). While Ross and Usher illustrate the northern Canadian process through describing the way of life in a remote Inuit community they also emphasise the existence of informal components in non-aboriginal economies, notably through family oriented small business, community organisations and other less rigidly structured groups. This sets their discussion within a broader development framework, in which the informal component of people's economic support systems has been progressively supplanted by the formal component. Their simple diagram (Figure 2.6) allows for changing emphasis between these two components, and encompasses the variability which more rigid development models have ignored.

The second model, the concept of the next economy, suggests another useful way of fostering remote area development through an approach which has been perceived as particularly appropriate for aboriginal communities (Robinson and Ghostkeeper, 1987: 138). The next economy is based on the principle that, at community level, economic activity should be small-scale, based on local physical and human resources and operated in such a way that the community supports it. In other words, the next economy promotes self-determination. In remote aboriginal communities, where natural resources are often scarce but human resources plentiful, the next economy would probably focus most appropriately on small enterprises or the service sector. The latter could be particularly important. In almost every aboriginal community in outback Australia or in Canada's north most jobs are in the service industries – in the schools, the health clinics, the shops, or the town council offices – and, with the lack of opportunities for growth in the industrial or manufacturing sector, it is in this area that aboriginal people should seek to increase their direct participation and control. This could be through taking over publicly funded service positions currently held by outsiders, often non-aboriginal people; and by developing 'next economy' opportunities, ones whereby these services are provided by local people under community control rather than externally by government agencies. The former approach presents problems. There will probably be a dearth of local aboriginal residents qualified to teach in the school, run the health clinic, or operate the power station. Moreover the qualifications demanded in such publicly funded positions can only be obtained through attending courses leading to professional certification, and few people will have sufficient formal education to meet the entry requirements. However such problems can be overcome. With careful planning and sufficient financial support aboriginal people can be given the necessary training and experience to meet these requirements. Moves to increase opportunities for management training, and to upgrade training and status for aboriginal teachers and health workers show that authorities are aware of the need (Ellanna et al., 1988).

The idea of the members of a community providing their own services has many attractions, particularly when the cultural and social life of that community are distinct from that of the majority population. Aboriginal people have already demonstrated the value of such an approach, through the establishment of their own schools and health clinics, or through setting up their own radio or television stations. These steps have not been easy. Commonly opposition from non-aborigines has been fierce: some are opposed because they see themselves losing their jobs; some, convinced that aboriginal people are incapable of organisation and administration, have refused their support on the grounds that failure is inevitable; and some, primarily bureaucrats, see it all as yet another waste of resources. As a result, many of these services have been forced to start on a shoestring, dependent on voluntary labour or on donations from individuals and from welfare organisations. This, as Robinson and Ghostkeeper (1987; 1988) point out, makes it essential that this component of the next economy has full community support, possibly including communal or co-operative owner-ship of the ventures. As they also point out, many of the attributes of activities in the next economy – small, family-based business or services, broad scope of skills, co-operation, self-reliance – accord closely with those found in the subsistence, 'informal' economy. These two models are therefore closely interlinked.

Ideas such as recognising and supporting the informal economy, or encouraging the next economy also acknowledge another important fact often ignored by both the private and public sector in remote areas. Simply stated, aboriginal peoples not only provide a high proportion of the population in those regions; their presence also supports a very significant proportion of the local economy. Their production, arts and crafts, furs and skins, and their labour in the pastoral industry, is the visible part of that contribution. The less visible component is the employment which they generate for others. Towns like Alice Springs and Yellowknife, with their predominantly non-aboriginal populations, owe much of their prosperity to their roles as service centres for aboriginal people. In Australia this kind of activity is often referred to as the Aboriginal industry. Its value to the local economy is high. In the late 1970s Drakakis-Smith (1984b) estimated that over one-third of the jobs in Alice Springs, the main town in Central Australia, were there by virtue of the Aboriginal industry. In a more detailed study Crough et al. (1989) confirmed this with calculations that the Aboriginal contribution to the total economy of central Australia in 1987/8, estimated at around $184 million, was about one-third of the total. Crough and Christopherson's more recent (1993b) study highlights a similar contribution by Aboriginal people to the Kimberley economy. Figures such as these should help to refute statements about aboriginal people contributing nothing except problems, and about them being principally a drain on resources. However such ideas remain heavily entrenched and, as Crough (1993) has much more recently pointed out, aboriginal people still remain largely invisible (i.e., ignored). They only become visible when their activities and ideas impinge directly on the interests of the non-aboriginal population. As indicated earlier, current moves for regional autonomy in remote areas are likely to provide such a focus.

Ideas like the 'informal' and 'next' economies are also compatible with the sustainable development model and it is this approach, combining economic, social and environmental elements in peoples' lives, which is the focus of this study. The definitions are purposely broad. Thus it considers economic contributions, including economic enterprises, as including all types of activity which sustain aboriginal people, whether or not these involve the use of cash, or whether or not they are commercially or subsistence oriented (Ellanna et al., 1988: 1). These are considered not on their own but in conjunction with the society in which they operate and against the background of the environment where the people concerned live. That combination is seen as essential for provision of an appropriate basis for achieving a better way of life based on self-determination. It can only occur if all three elements are taken into account in support given to aboriginal groups. That is often lacking. Thus, in the social sphere, if education is offered solely within the non-aboriginal mould aboriginal children will not learn the skills which they need to cope with the broad-ranging nature of their society; similarly, in the environmental system, governments may offer aboriginal communities funds to conserve en-dangered animals (the environmental element) but they may fail to recognise that people in that community value these animals more highly for food (the economic element); not surprisingly their efforts will fail. And in economic terms, we need point only to the problems which arise when community-based enterprises are judged solely from the perspective of their commercial viability, with no account taken of their social contributions. Any of these examples could be used to answer the question of why aboriginal development has so often failed. As the next chapter, which examines government development policies and practices towards remote-dwelling aboriginal Canadians and Australians, suggests, the problem has often started at government level.

The development potential of remote-dwelling aboriginal groups

As this brief overview indicates, aboriginal people in remote Canada and Australia are disadvantaged on any conventional non-aboriginal measure. They are uneducated and poor and they live in extreme isolation compared to the core regions of their countries. They lack the financial resources to buy into the system, and even to buy the expertise which might help them to overcome their own lack of skills. Many are unemployed and, with the increasingly high levels of technology required in most industries today, their chances of obtaining paid employment are minimal. True, they now hold significant stretches of land. Khalidi's recent comparison of socio-economic status and land holdings shows that the most dis-advantaged regions of Australia are also the areas with large stretches of Aboriginal land (ATSIC, 1992a). However most of that land is resource poor. This scenario is not only depressing; it is also one which appears to offer little potential for improvement. However such an assessment would be unduly pessimistic. The situation of aboriginal people also has to be seen from their perspective, one in which the regaining of the land, with its immense spiritual and cultural value and its wealth of renewable resources, plays a central part. These, along with their knowledge of survival in such difficult and harsh environments, are the assets which aboriginal peoples bring to development. They are assets which have frequently been ignored. A prime question is whether development policy, as evolved and currently practised by Canadian and Australian govern-ments and by private operators, is able to accept the challenge and adjust to the opportunities which it presents. However they are, and will probably remain, heavily dependent on support through the government (public sector). The quantity and quality of that support, coupled with the way in which aboriginal people respond to the opportunities made available to them, provide the key to their development.

3

GOVERNMENT POLICIES AND PROGRAMS FOR ABORIGINAL DEVELOPMENT

Aboriginal people in Canada and Australia are, of course, free, like any other citizens, to initiate and support their own development plans and projects. In practice their socio-economic disadvantages and lack of access to resources or to the capital to exploit these make it difficult for many of them to take such steps. Government development programs are therefore of over-whelming importance, particularly for people in the more remote parts of the country. The federal governments in Canada and Australia, through the Department of Indian and Northern Development (DIAND) and the Aboriginal and Torres Strait Islander Commission (ATSIC) respectively, take overall responsibility for the affairs of their aboriginal peoples. Development policies and programs designed to implement those policies are an integral part of that responsibility. Provincial and territory governments are also involved, both through their local administration of programs emanating from DIAND and ATSIC and also, to a much lesser extent, through their own programs, which apply to aboriginal as well as non-aboriginal applicants.

In both countries present day policies affecting aboriginal development reflect a history of changing attitudes, including colonialist and assimilationist views and more recent acceptance of the need for aboriginal self-determination coupled with greater economic self-sufficiency and community-based development. In general, as Cant's recent (1993) comparative overview of the basis for land rights movements and development in Canada, Australia and New Zealand sets out, there has been marked progress in the movement from assimilation to self-determination since the late 1960s. However the transition has not been smooth. Many Canadians and Australians still openly espouse the cause of assimilation and condemn government for supporting aboriginal cultural and social development, and for making any special concessions to processes to overcome aboriginal economic disadvantage. They take the attitude that those who cannot or will not join the mainstream do not deserve to survive. Such views are also common with those involved in government, both among the politicians and amongst the bureaucrats. In Australia, for example, despite acceptance that Aboriginal

89

Affairs will as far as possible be dealt with not along party lines but with a bipartisan approach, many Aborigines are convinced that right-wing political factions will deal harshly with the portfolio, espousing assimilation under the more modern rubric of 'mainstreaming'. The survival of assimilation alongside the current policies of self-determination leads to great uncertainty and has been a source of enormous frustration for aboriginal peoples. Current policies and programs can only be understood against this background.

CANADIAN FEDERAL POLICIES AND PROGRAMS

The administration of Indian Affairs in Canada, at various times within the Departments of Mines and Resources, Northern Affairs and National Resources and, since 1966, the responsibility of DIAND, has generally emphasised a welfare rather than development policy. This reflects its establishment in response to the destitution of Indians displaced from their lands, a situation which demonstrated the wrongful assumptions from the original treaties that land would be exchanged for an economic base which would allow for self-determination. Loss of land and population growth meant that this balance was impossible to maintain by the mid-twentieth century and Indians were then forced to rely on government hand-outs. By the 1960s, however, aboriginal people and others were strongly criticising both the welfare emphasis of DIAND's policy and the paternalistic nature of its administration. Elias (1991) attributes this to a number of factors, including the return of aboriginal members of the Canadian armed forces after the war and other questions raised by third world decolonisation, the American civil rights movement and student and youth activism. These criticisms subsequently led to a series of reports and reviews which together form the basis for current aboriginal development programs.

An early response from DIAND was a study of the contemporary economic, political and educational needs of Indians, reported upon by Hawthorn (1966–7). This report, the first contemporary study of Indian conditions, exposed their dismal situation as the most poverty-stricken group in Canada and saw the solutions to this not in assimilation but rather in improving their chances of sharing, on their own terms, in the opportunities offered to all Canadians. DIAND, while this study was still incomplete, began to make their own moves, decentralising some of their responsibilities, including education, to the provinces and beginning to implement community development programs. Despite these responses welfare programs continued to dominate their spending, with a 1968/9 budget which was seven times that of the allocation for economic development, and as Weaver (1981: 84) has said, DIAND continued to maintain its custodial role.

In 1968 Cabinet commissioned a policy review of DIAND which resulted in the White Paper of 1969. The findings of this paper were universally rejected by aboriginal Canadians and also by many top-level public servants concerned with aboriginal affairs. The White Paper effectively recommended the abolition of the Indian Act and therefore of DIAND's Indian Affairs branch, and failed to address broad spectrum issues, including the need for land rights and an approach to development which was not solely based on economic advancement but which also incorporated social change. Aboriginal people also complained about the government's lack of consultation with them and the secrecy surrounding the whole review process. Criticism was so severe that in 1971 the paper was formally withdrawn.

Dissatisfaction with the 1969 White Paper had, according to Weaver (1981: 5), an important spin-off – it led to an upsurge of aboriginal nationalism and determination to exert pressure for recognition of aboriginal views of development. In the early 1970s reports submitted by the Manitoba Indian Brotherhood, the Indian Chiefs of Alberta and the Council for Yukon Indians all stressed the importance of comprehensive approaches to development, including land rights and greater control over resources. They also pointed out that, although many aboriginal groups saw the importance of economic aspects of development, socio-cultural and environmental issues might well be the first priority. As Elias (1991: 15), suggests, this view did not accord with that commonly held by the federal government.

During the remainder of the 1970s and the early 1980s the focus was on bringing aboriginal and government views of development together, in an attempt to reach a workable compromise. This proved to be a difficult process. In 1976 a committee was formed under the joint auspices of DIAND and the National Indian Brotherhood (NIB) to consider Indian socio-economic development strategies. In a 1977 report this committee confirmed the views of those aboriginal organisations which had already responded to the White Paper, rejecting assimilation and stressing the importance of holistic views of development. They did, however, expect this to occur within the framework of existing Canadian institutions, a restriction which did not accord with more radical aboriginal views which wanted revision of the whole system. Local self-government, based on the devolution of administration from DIAND to band governments, was seen as the appropriate avenue. Another joint committee, also under DIAND and NIB auspices, examined DIAND's role and operations but, because of subsequent differences of opinion, never produced a final statement. An interim report (Beaver, 1979) made some useful comments on the reasons for DIAND's failure to advance development. According to Beaver any development being promoted by DIAND was the outcome of DIAND's rather than aboriginal aspirations, and there had been a failure to understand that development was a process of evolution and not something that 'could be ordered [or] bought [by officials of Indian Affairs]' (Beaver, 1979: 29). Beaver also commented on the overwhelming emphasis placed by DIAND on welfare or remedial programs rather than on development. Local self-government at community level was seen as the prime method of reversing DIAND's 'top-down' development approach.

Aboriginal organisations, wary of recommendations which saw aboriginal development fitting into the Canadian institutional framework, did not wholeheartedly support Beaver's comments. Nevertheless many of his points subsequently formed the basis for ongoing discussions with DIAND. Penner's report on the establishment of Indian self-government (House of Commons, 1983), the result of a House of Representatives task force review on Indian self-government, emphasised the same points. This report also made important comments on the need for economic foundations to support Indian controlled self-government, highlighting the following key elements to be transferred from DIAND: land/resources; capital; labour; organisation; planning; and technology. Amongst other things the report pointed out that DIAND's retention of control over land undermined Indian development potential because bands could not raise capital by using their land as security. Moreover the taxation concessions granted through the Indian Act for revenues raised on reserves disappeared if these were raised by Indian controlled enterprises incorporated in order to qualify for funding.

Beaver's and Penner's reports essentially said the same: that aboriginal development should be promoted through community-based self-government with a sound and independent economic base; and that this could only be accomplished through the reform of DIAND's operations. Aboriginal Canadians were optimistic that these findings would play a prominent role in future federal development policies. They were to be speedily disillusioned. In 1984/5 the newly elected conservative government commissioned a task force review, under the chairmanship of Erik Nielsen, the Deputy Prime Minister, to consider cost-cutting measures for all government programs including Indian and Native programs (DIAND, 1985 and 1986). Despite clear statements from Prime Minister Mulroney at the opening of the First Ministers Conference (FMC) in 1985 that the goal of native self-government would be pushed as the solution to poverty and dependence, preliminary findings from the Nielsen task force, leaked shortly afterwards, indicated that cabinet thinking did not accord with such a line (Weaver, 1986a: 4). The duplicity implied by such evidence, and the secrecy with which it had been compiled, reminded many people, native and non-native alike, of the events surrounding the White Paper of 1969. The Nielsen Report recommended transferring DIAND's existing economic and employment programs to the Department of Regional and Industrial Expansion, now the department of Industry, Science and Technology Canada (ISTC) and to Canada Employment and Immigration Commission (CEIC), and the transfer of Indian programs to provincial responsibility. It placed strong emphasis on economic rationalism, emphasising that programs would operate strictly on a commercial basis and that small business would play an important role; and that service delivery programs would be run on a user pays system, with no increase of funds for the foreseeable future (Deputy Prime Minister's Office, 1986). Neither consultation with aboriginal people nor self-government and community development were mentioned.

This report was fiercely rejected by Canada's aboriginal people, not only because of its failure to consider earlier findings, but also because they were apprehensive about plans to dismantle DIAND. They saw this as a deliberate move to deny them their unique rights and heritage and treat them the same as other poorer Canadian citizens. Like the 1969 White Paper, the Nielsen report was also rejected by government. However it has since been noted that many of the directions stressed in the report have been followed. Erasmus (1986: 56–60) comments that subsequent community-based self-government arrangements (opting for municipal-like government not necessarily under the Indian Act rather than self-government as desired by native peoples); Canada jobs strategy (emphasising the private sector and encouraging people to leave their bands to seek work elsewhere); alternate funding arrangements (stressing greater accountability at band level for spending of federal funds); economic development (stressing individual entrepreneurship rather than band controlled enterprises) contain many elements from Nielsen's recommendations.

There are a number of important interlinked threads running through this recent history of Canadian policy affecting aboriginal development. First, while the watchwords of self-determination, self-government and community government continually appear, the actual definition of these terms is by no means consistent. Aboriginal understanding of self-government, for example, implies governing at community level with appropriate resources to do so effectively and the power to decide how such resources should be used; the bureaucratic understanding is that government will retain financial control and will attempt to enforce a common development model on all groups. Aboriginal organisations, such as the Assembly of First Nations (AFN) later made the same point; that under pressure the term 'community' was invariably deleted from development definitions and that the concept of commercial viability was always seen as a priority (Dedam, pers. comm., 1989). Secondly, economic development has generally been defined restrictively, stressing monetary profit-making and ignoring both the non-monetary sector and the social context of economic activity. DIAND officials know about these problems. However the economic circumstances of the late 1980s to early 1990s have made them particularly sensitive to requirements for strict accounting for expenditure by measurable criteria, and they have been unwilling to incorporate less easily assessed social benefits to justify their activities (Kariya, pers. comm., 1989). As this suggests, acceptance of a policy of self-determination does not mean that such a policy will be practically realised. The following summary of the Canadian government's recent aboriginal development programs makes this even more obvious.

Table 3.1 Canada, Commonwealth aboriginal economic development programs

Departments	Program	Characteristics
DIAND	IEDF ELF ICHRS IOGC Resource development	Loans for business development Loans for business development Training in economic/community development Oil and gas development Renewable resource development
	Special ARDA/NDA (to 1989)	Grants for income, employment in remote regions (aboriginal and non-aboriginal populations)
ISTC	NEDP (1984–9)	Element I – Support for Native Capital Corporations (NCCs) (loans) Element II – Support for community-based economic development Element III – Loans for commercially viable enterprises
DIAND *ISTC* } *CAEDS* *CEIC* } *(1989→)*	Business development Joint ventures Capital corporations Community economic planning and development Access to resources Skills development Urban employment Research and advocacy	Loans for commercially viable business ventures: loan guarantees Funding assistance for aboriginal participation in joint ventures Support for maintenance and expansion of NCCs Support for human resource development and advisory bodies for economic planning at community level Financial assistance for aboriginal groups to share in business opportunities in the resource section Training for business skills Training specifically focused on the urban unemployed and underemployed Research on aboriginal enterprise performance, and dissemination of information to the public

Recent and contemporary programs for aboriginal economic development in Canada

Funding for aboriginal economic development in remote parts of Canada has, for most of the period since the 1970s, come primarily from two federal government departments, DIAND and ISTC. A third department, CEIC, is involved in the current program, CAEDS. Table 3.1 summarises the main characteristics of the most important programs.

DIAND

DIAND's contribution to aboriginal economic development, although accounting for only 4.4 per cent of its total Indian and Inuit Affairs budget in 1987/8, has been important, particularly in remote areas where few alternative avenues for support have existed. Its programs can be categorised as those concerned solely with economic development based on business projects which are commercially viable – the Indian Economic Development Fund (IEDF) and the Eskimo Loan Fund (ELF); those that are concerned with human resource development, primarily the provision of training for economic development management – the Indian Community Human Resource Strategies Program (ICHRS); and those concerned with providing support for aboriginal Canadians to share in resource development, both renewable and non-renewable – Indian Oil and Gas Canada (IOGC) and other minerals and resource development.

IEDF and ELF, seen as programs to assist aboriginal people to set up 'viable businesses which contribute to wealth creation, the enhancement of employment opportunities and to economic independence' (DIAND, 1989b: 2–40), have since the 1970s offered direct loans and loan guarantees and also advisory and financial services. In 1986/7 this program accounted for 27.6 per cent of the economic development expenditure and 964 projects were receiving support. ICHRS is designed to provide support so that communities can 'organize long-term human resources development strategies' (i.e. planning) to complement other components, for example business development, through activities such as business management training, career counselling and youth entrepreneurship. It accounted for the highest proportion of the economic development expenditure in 1986/7, 38.8 per cent. During that year ICHRS funded 230 economic development officers working at band level, as well as almost 3,500 trainees working in various areas to improve their skills. IOGC 'provides professional advice and assistance to Indian and Inuit people in identifying, developing and producing mineral resources' and resource development provided similar support to projects in the renewable resource area, particularly tourism, agriculture, fisheries, forestry and wild fur management. Together they accounted for about 29 per cent of DIAND's economic development expenditure in 1986/7.

A major problem affecting DIAND's economic development programs has been their stress on loans rather than grants. The IEDF has, ever since it was set up in the early 1970s, had great difficulty in persuading people to repay their loans and there have been suggestions that it should be wound up. However it has also been acknowledged that, if it had not existed, many small aboriginal businesses would never have started and that its function in securing bank loans for applicants has been particularly useful (DPA, 1985). Suggestions that aboriginal people, through their native development cor-porations, should take over responsibility for managing these funds have more recently been supported by DIAND. Kariya (pers. comm., 1989) suggested that since this might ensure that the funds would be used for loan guarantees rather than direct loans more aboriginal people might then be able to access money through conventional banking systems.

General difficulties faced by Indian businesses included lack of access to conventional banking institutions; lack of capital/equity to enable them to compete to successfully apply for contracts even in their own communities; lack of joint ventures with non-aboriginal businesses, a strategy which might have improved their success rate; lack of native management skills; and the need to encourage community-based business development, satisfying local markets and using local resources, both renewable and non-renewable (DIAND, 1986). An additional problem was that DIAND's programs were poorly co-ordinated with those offered by other agencies at both federal and provincial levels.

ISTC

ISTC and its predecessor DRIE have offered two programs which have supported remote area aboriginal economic development – Special Agricul-tural and Rural Development Agreements (Special ARDA) and Northern Development Agreements (NDA); and the Native Economic Development Program (NEDP). The agreements, of which Special ARDA includes the Northwest and Yukon Territories, stemmed from the 1970s and were designed particularly 'to improve the income and employment opportunities of disadvantaged people, particularly those of native ancestry, living in rural and remote areas' (DIAND, 1989b: 21). Special ARDA programs have not only supported commercial enterprises but (Table 3.2) have also met some human resource needs. Primary production, including assistance for the improvement of the efficiency of hunting and gathering activities, received significant support in Manitoba and NWT. Altogether, while the three prairie provinces have been the main beneficiaries from Special ARDA, the scheme has made an impact in the remote north, with over $27 billion dollars going into projects in that region between 1982 and 1989 (ISTC, 1989a).

Unlike Special ARDA, NEDP, a national program launched in 1983 with initial funding of $345 million over four years (DRIE, 1984), was specifically aimed at supporting aboriginal development. It terminated in 1989. Its objectives were as follows:

i) to increase and strengthen aboriginal community projects that have a strong economic focus, that increase the self-reliance of aboriginal people, and that have the potential to be commercially successful

ii) to increase the number of aboriginally owned and controlled enterprises, including financial and economic institutions and businesses

iii) to increase the access of aboriginal people to existing economic development resources in the public and private sector

iv) to increase public awareness of the contribution to the mainstream economy by aboriginally owned, managed and directed enterprises.

(DIAND, 1989b: 20)

The program contained four main elements, of which Element I, supporting the establishment of aboriginally controlled and owned capital corporations, and Element III, enterprise support, received the bulk of the funding.

By March 1989 NEDP had helped to establish 26 aboriginal (native) capital corporations (NCC) throughout Canada, 17 of which were already operating, 12 with loan portfolios exceeding $1million. They aimed to provide loans at commercial rates for aboriginal clients, most of whom would otherwise have difficulty in obtaining finance. While a final assessment of their contribution was not then available, an earlier review (DRIE, 1988) indicates some characteristics of their operations. In 1988 15 NCCs supported by NEDP were operational, all with native controlled boards and some with native managers. They had disbursed $36.4 million in loans, of which 34.1 per cent had been fully repaid; 87 per cent of borrowers were satisfied with the service they had received; and people's understanding of how such commercial institutions operated was already being enhanced. While some NCCs (for example, NWT Co-op Corporation, allied to Arctic Co-ops Ltd (see Chapter 7)) had developed from existing operations, others had been spawned by aboriginal political organisations, an interesting situation because of the possibility of conflicting commercial and political interests. Borrowers from NCCs would not have been attractive clients to conventional banks – their needs were comparatively small, and therefore expensive to administer, and they would have had difficulty in persuading banks to accept them.

NEDP initially assumed that NCCs would themselves eventually operate as bank-like institutions, not only granting loans but also accepting deposits and investing in property and the money market. In mid-1988, still the establishment phase for most NCCs, this was yet to occur. It was also suggested that in the long term they might take over the functions of other government economic development programs, such as IEDF, a possibility which would certainly concur with the aim of self-determination. The fine line between generating profits and providing a service to borrowers, particularly problematic because of the small size and isolated nature of native enterprises and the management difficulties they often encountered, was also perceived to be an issue. NEDP's job was difficult. As a public funding body it had to guard against misuse of government monies and NCCs were firmly told that, if necessary, support would be withdrawn and they might be allowed to go bankrupt. At the same time NEDP would let each NCC control its own operations, and would confine its own activities mainly to keeping records of its financial dealings with each corporation.

By 1989 NEDP had assisted over 400 aboriginal commercial enterprises, accounting for approximately 50 per cent of their equity. As Table 3.2 shows, manufacturing, agricultural and wholesale/retail enterprises were the most important categories and the main regions receiving benefits were the provinces of British Columbia, Alberta and Ontario. In relative terms the two northern territories and the maritime provinces had fared badly. Not surprisingly, more than half of the NWT projects which had received NEDP assistance belonged to the agricultural (which included fishing, trapping and forestry) and tourist categories (ISTC, 1989a and 1989b).

A major reason for remote Canada's low participation rate in NEDP's enterprise program was the emphasis placed on commercial viability. Success in loan applications tended to be based on conventional indicators which measured material wealth; being able to provide a high level of equity, or having a proven commercial record. Indicators that enterprises would benefit community development were not highly valued. This issue was mentioned in criticisms made by aboriginal Canadians. Other concerns were the lack of co-ordination between ISTC's programs and those offered by other government departments; the need for greater program flexibility in order to cater for the diversity of situation in aboriginal communities; the need for training and for better support for program participants; the need to decentralise ISTC's activities from major metropolitan centres such as Winnipeg, Vancouver and Montreal; and the need to have a majority of aboriginal people on all assessment boards (DRIE, 1988). Additional issues considered were the effectiveness of consultation processes involving the establishment and implementation of NEDP (ISTC, 1988) and the limited involvement of women in NEDP supported programs (NEDP, 1985).

Table 3.2 Special ARDA and NEDP, Element III

	NEDP																Special ARDA		NEDP and spec. ARDA % share by region
	Agric., forest, fishing		Mining		Manufacture		Construction/ transport		Wholesale/ retail		Accomm., food		Other		Total		NEDP % share by region		
	Amt. $'000	%	Amt. $'000	%	Amt. $'000	%	Amt. $'000	%	Amt. $'000	%	Amt. $'000	%	Amt. $'000	%	Amt. $'000	%		Amt. $000	% share by region	
E. Canada (Ont., Que., Newf., Maritimes)	2,744	7.5	374	1.0	5,053	13.7	6,432	17.5	5,844	15.9	5,725	15.6	10,580	28.8	36,754	100	29.5	–	–	15.2
Prairies (Sask., Man., Alberta)	4,095	9.3	3,088	7.0	6,077	13.8	6,702	15.2	3,614	8.2	5,596	12.7	15,017	34.0	44,187	100	35.4	63,207	53.9	44.4
British Columbia	4,223	13.0	–	–	10,021	30.9	6,698	20.6	1,532	4.7	2,290	7.1	7,709	3.7	32,473	100	26.0	27,050	23.1	24.6
North (NWT, Yukon)	2,119	18.8	–	–	1,063	9.5	1,430	12.7	3,151	28.0	2,297	20.4	1,188	10.6	11,248	100	9.0	27,000	23.0	15.8
Total	13,181	10.6	3,462	2.8	22,214	17.8	21,262	17.0	14,141	11.3	15,908	12.8	34,494	27.7	124,662	100	100	117,257	100	100

Source: Compiled from Gibson, 1989.

Canadian Aboriginal Economic Development Strategy (CAEDS)

Both DIAND and ISTC had isolated some key problems affecting their delivery of aboriginal economic development programs – their failure to co-ordinate with each other to prevent overlap and to ensure that they used each others' strengths to provide a better service; and the short-term nature of much of their support. While this suggested that programs should be streamlined, such suggestions were not universally welcomed because each department was worried about the loss of any part of its portfolio and the resultant threat to jobs. Following an interdepartmental review at the end of NEDP's initial period, a new strategy, the Canadian Aboriginal Economic Development Strategy (CAEDS), was launched. This strategy brought together the aboriginal economic development activities of three federal departments – DIAND, ISTC and CEIC. Aimed at promoting 'the long-term employment and business opportunities of Canada's aboriginal citizens, by giving them the means to manage effectively their own business enterprises, economic institutions, job training and skill development', it was funded for the first five years with a total of $873.7 million, more than half of which came from DIAND (Canada, 1989: 5).

CAEDS had eight elements (Table 3.1); some were designed to suit commercially oriented projects (for example, business development); some, while aimed at economic development, were set within a context which would take account of particular problems faced by aboriginal people (for example, joint ventures and capital corporations); some supported human resource development (for example, skills development); and some were designed to bring economic and social aspects development together (for example, community economic planning and development).

Each department took responsibility for whatever program fitted most closely to their existing skills and resources. Thus ISTC took business development, joint ventures and capital corporations, DIAND took community economic planning and access to resources, CEIC took skills development and urban employment; and the responsibility for research and advocacy was shared. Table 3.1 gives further details on what each program covered. Those areas likely to be most relevant to the needs of people in remote Canada included the micro-enterprise development component of business funding, which was designed to deal with the capital needs of aboriginal people with little or no commercial experience working on a small scale. It is in this section that ventures associated with hunting and foraging might well fall. Joint ventures, which are likely to enhance opportunities for resource development, including mining; and access to resources, which would assist aboriginal groups to make applications for participation in resource development, would also be relevant in Canada's north.

The structure of CAEDS theoretically reflected long and detailed consultations between members of the federal departments concerned and also between them and Canada's aboriginal population, and it was assumed that strong support would be given. However communication even in the early stages seemed to have been less successful than expected. In the absence of a detailed review of the subsequent operation of CAEDS, the comments of those concerned about the strategy in late 1989 raise some interesting questions.

Elias (1991: 33) has stated that, although CAEDS appeared to accord more closely with aboriginal development ideals, it did so within a very strong focus on economic advancement. In late 1989 economic development officers working with two aboriginal organisations, the Assembly of First Nations (AFN) and Inuit Tapirisat Canada (ITC) made similar comments. They were also sceptical about whether these federal departments could ever work together; whether CAEDS funding would primarily assist aboriginal development projects, or whether it would find its way into the hands of white entrepreneurs; whether social and cultural attitudes and attributes would be recognised as valid elements in applications; and whether consultation between the government departments and native organisations had been and would continue to be adequate.

AFN were also concerned that CEIC, a vital part of the strategy because of their role in human resource development, seemed to be less committed to CAEDS as a whole. AFN also felt that DIAND would be quite incapable of shedding its colonial mentality, and that it would continue to hold on to control. From their observations at joint meetings between all parties, they felt that DIAND seemed to have made prior decisions about the expenditure of funds, and that collective decisions might well be pre-empted. DIAND's use of their existing regional structure to disseminate information and initiate applications from CAEDS (through their band economic development officers) was also seen as a potential problem because it allowed for continual entrenchment of the 'paternalistic' status quo. AFN felt more optimistic about ISTC, perhaps because their experience with this department had been more limited but also because ISTC had previously been so heavily centralised. As they suggested, this might make their attitudes more flexible and increase the chance for CAEDS to have a structure which accorded directly with native people's ideas (Dedam, pers. comm., 1989).

ITC, the main Inuit organisation concerned with economic development, stressed their concern that CAEDS funds might be diverted from aboriginal entrepreneurs, particularly through joint ventures. They commented that aboriginal clients were often told that they had to agree to joint venture arrangements if their applications were to succeed; but that their non-aboriginal partners were not necessarily willing to keep their part of the bargain, particularly in management training and the transfer of other skills. Their economic development officer commented (Higgins, pers. comm., 1989) that there were a number of such situations where, instead of being the best person for working on the project, the non-aboriginal partner was a 'mate' of local government officials.

101

Both organisations also questioned whether CAEDS would provide any support for projects involving traditional subsistence activity. They felt that these could be particularly important for increasing economic self-sufficiency in more remote communities. They perceived that, because departments like DIAND still had problems in accepting that cash was an essential part of a viable subsistence economy, there would be major difficulties in obtaining funding for such projects as, for example, fishing allied to tourist game fishing and outfitting (Dedam, pers. comm., 1989). These concerns may well have been justified. As Elias (1991: 32) notes, programs to assist subsistence activities would come under community development, for which no detailed plans were announced until 1991. Even these plans, as he further comments, emphasised long-term economic benefit as a major goal. Both organisations were also generally sceptical about whether these departments were capable of incorporating social elements into their assessments for support.

Finally, while both AFN and ITC said that their consultation with the three departments had worked reasonably well in Ottawa, they were concerned about communication elsewhere. By mid-October 1989, six weeks after the program was supposed to be operational, application forms and information sheets for bands and communities had only just become available, and there had been no opportunity for these groups to plan their responses. ITC, in an attempt to counteract this problem, were preparing to ask DIAND for substantial support from the human resources component of CAEDS to set up an Inuit planning agency specifically designed to help individual Inuit applicants to successfully apply for enterprise support.

The three departments involved in CAEDS had also encountered difficulties in communication. Both DIAND and ISTC lacked confidence in CEIC's commitment to the idea that special programs for the assistance of aboriginal people were necessary. CEIC have always emphasised the equality of status of all Canadians, regardless of ethnic origins; they have also relied on clients contacting them, without them making any special effort to disseminate information (CEIC, 1989). Their high level of decentralisation, with a much larger number and geographical spread of local offices than those run by either DIAND or ISTC, has helped them to maintain this stance. There was also some overlap in the existing activities of the three departments, although solutions were being proposed. CEIC anticipated that it would work primarily with the urban areas, the regions less commonly dealt with by DIAND; and it felt that its business development centres could effectively complement the NCCs rather than conflicting with them.

Australian federal policies and programs

Prior to the 1967 referendum, when Australia began the transition to self-determination, Aboriginal affairs were administered under assimilation policies. State governments, then responsible for Aboriginal administration within their own regions, and the Northern Territory, under the federal government, operated under very similar guidelines. Welfare, rather than development, was the emphasis, and paternalistic attitudes prevailed throughout all Aboriginal affairs bureaucracies. Development under assimilation was perceived as a process through which Aborigines would learn to contribute to material advancement for both local and national good, in the process denying their language, customary traditions and social structure. Economic development elements included encouraging and/or coercing all Aborigines to accept European work ethics and join the labour force, primarily in positions at the bottom end of the occupational hierarchy. It was generally assumed that few would be able to take on responsible or supervisory roles. Rewards, initially in kind rather than in cash, were small and Aboriginal workers were instructed and supervised so that, according to non-aboriginal perceptions, they spent their rewards wisely. In practical terms such policies have had long-lasting effects on Aboriginal attitudes towards and experience of participation in the wage/cash economy.

From the early 1960s the situation began to change. As in Canada the growth of an Aboriginal voice in matters affecting their own destiny, and the recognition by administrators that, certainly in the more remote regions, the assimilation policy was not having the desired effect helped to influence attitudes. Restrictions gradually relaxed in the early 1960s included those affecting liquor consumption and, following the withdrawal of the Welfare Ordinance for the NT in 1964, rules affecting elements such as Aboriginal mobility and rights of residence. In 1966 the first overt demonstration of Aboriginal land rights aspirations in the remote north occurred with the walk-off of the Aboriginal workforce in Wave Hill cattle station and the subsequent establishment of the Gurindji settlement at Wattie Creek. Finally in 1967 came the referendum.

The referendum, which gave overall responsibility for Aboriginal Affairs to the federal government, theoretically opened the way for a revision of Aboriginal development policy, one which took account of the frustrations of Aboriginal people, and their clear desires for cultural and political autonomy allied to land rights. This new policy was tentatively referred to as integration, a process by which Aborigines would become part of the majority Australian society without losing their language, their identity and their cultural traditions. In this way their transition would be more akin to that of present day non-Anglo-Saxon migrants to Australia, a component in the development of multiculturalism. Economic development would obviously form a vital part of integration. Such a change would require not only the adoption of such a policy at Commonwealth level, but also the assumption by the Commonwealth of its legal responsibilities which would allow state rules to be overridden if they did not accord. Change was slow. As Rowley (1978: 18) pointed out, Australian Aboriginal policy had been 'a by-product of policies related to economic development for whites – policies about land, labour, migration, mining and trade'. Such thinking was entrenched and as a result the federal government was slow to exercise its powers, particularly in areas of potential conflict. Thus, when Aboriginal pastoral workers were finally granted the Award wage in 1966 the decision was not implemented for another three years so that Europeans in the cattle industry would have time to adjust to the new employment situation.

Following the referendum the federal government established a Council for Aboriginal Affairs, under the chairmanship of Dr H.C. Coombs, former governor of the Reserve Bank; and set up an Office for Aboriginal Affairs within the Prime Minister's Department. As Coombs (1978) has documented, members of the Council, essentially an advisory body, became more and more strongly convinced that assimilation should be discarded. Their view was strongly supported by the findings from the first major study of the socio-economic condition of Aborigines, carried out during the second half of the 1960s under the auspices of the Academy of Social Sciences and directed by Dr C.D. Rowley (see, for example, Rowley, 1970a, 1970b and 1970c; Gale, 1972), findings which essentially showed that the policy had failed. In contrast bureaucrats in the Department of the Interior continued to support assimilation. However, as Prime Minister McMahon's statement in 1971 – that Aborigines should have the right to determine the pace and direction of their future development – shows, there was support for self-determination at the head of government. Others still failed to accept this, and in early 1972 a parliamentary Committee on Aboriginal Affairs announced a new policy which ignored the advice of the Council and wholly supported assimilationism. Strong Aboriginal opposition to this led to the setting up of the Tent Embassy on the lawns outside Parliament House in Canberra, and the government, embarrassed by the widespread publicity arising from this demonstration, at last opened channels for communication and consultation. At the end of that year the Australian Labor Party (ALP) came to power on a strong policy of Aboriginal self-determination, including land rights, a land purchase fund, anti-discrimination laws, legal aid and the establishment of a statutory basis for Aboriginal communities and their organisations.

The ALP was quick to keep its word. Aboriginal Affairs was upgraded to full departmental status, as the Department of Aboriginal Affairs (DAA); Mr Justice Woodward was appointed to establish how land rights legislation could best be implemented in the Northern Territory, the only area already under direct Commonwealth control; and an Aboriginal Land Fund Commission (ALFC) and Aboriginal Loans Commission (ALC) were established. Although not all of these moves, particularly that concerning land rights in the Northern Territory, were complete when the ALP lost power in 1975, the incoming coalition government continued the policy of consultation and the remaining legislation was passed with little revision. These elements, all part of a framework designed to implement self-determination, formed the structure of government policy throughout the remainder of the 1970s.

In 1980 the administrative structure changed once more with the formation of an agency concerned primarily with economic development: the Aboriginal Development Commission (ADC), which combined the roles of ALFC, ALC and the enterprise support section of DAA. The purpose of the Act which established the ADC, with a membership of ten Aboriginal Commissioners appointed by the Governor General, was:

> To further the economic and social development of people of the Aboriginal race of Australia and ... in particular [as a recognition of the past dispossession and dispersal of such people] to establish a Capital Account with the object of promoting their development, self-management and self-sufficiency.
>
> (Sect. 3, ADC Act, 1980; quoted in ADC, 1981)

Its philosophy aimed 'to maximise the well-being, self-determination and self-management [of Aborigines] so that they can enjoy a level of economic and social opportunity equal to that of other Australians' (ADC, 1986: 3). ADC's priorities in terms of the economic and social benefits of development have been the subject of conjecture ever since its establishment. In its early stages it was firmly stated of the projects it supported that 'Ventures need not necessarily be commercially viable [if] they can contribute to the social *or* [my emphasis] economic benefit of the community involved funds may be forthcoming.' (ADC, 1981: 9). The stress on economic and/or social independence gained through enterprise development was seen to be particularly important in remote areas, where the establishment of such projects not only provided much needed employment but also might conceivably save the government money in unemployment benefits and other welfare. Another important issue concerned ADC's status as an Aboriginal organisation, on the basis of an all-Aboriginal commission as the decision-making body. This body functioned within the bureaucratic structure of government, and many Aboriginal people have been sceptical about the reality of Aboriginal control exerted within it. Since most of the senior public servants responsible to ADC were ex-DAA staff, and in many cases ex-Northern Territory Department of Welfare staff, entrenched assimilationist ideals persisted and it was difficult for Commissioners to assert themselves against the pressures which they faced.

Yet another change took place in early 1990, when DAA and ADC were finally amalgamated to form the Aboriginal and Torres Strait Islander Commission (ATSIC). This body was the outcome of over two years of discussions and negotiations between the Minister for Aboriginal Affairs and Aboriginal organisations and communities throughout Australia. According to its Corporate Plan, 1992–6:

The goal of the Aboriginal and Torres Strait Islander Commission is to secure the empowerment of our people so that, through self-determination, we can make the decisions that affect our lives and share in Australia's land, wealth and resources, contributing equitably to the nation's economic, social and political life, with full recognition of our indigenous cultural heritage as the first Australians.

(ATSIC, 1992b)

ATSIC advises the Minister for Aboriginal Affairs on all matters concerning the interests of Aboriginal and Torres Strait Islanders groups. In contrast to both DAA and ADC, ATSIC's prime decision-making body is its board of 20 Aboriginal commissioners, 3 of whom are appointed by the Minister and the remainder of whom are the elected representatives of the 17 zones into which the country is divided.

ATSIC's goal is implemented through a wide range of programs grouped into three main categories: economic, including commercial, employment, development, education and training and regional support programs; social, including land, heritage and environment, health, social justice and infra-structure; and corporate services, including ATSIC's management, its offices of evaluation and audit and of public affairs, and its corporate services division. As these programs suggest, it retains the strong social and community development role, formerly under DAA; and also, through business funding, administration of the Aboriginal Employment Develop-ment Policy (AEDP), programs in land acquisition and management and the ATSIC Development Corporation (ATSICDC), it promotes economic development.

ATSIC aims to give Aboriginal people a greater say in the expenditure of Commonwealth funds allocated for their benefit. Its structure, with sixty regional councils of elected Aboriginal members each of which is responsible for determining the priorities for the expenditure of Commonwealth monies within its region, is intended to provide the federal government with a regional input of views much more broadly based than previously occurred. While this falls far short of the political autonomy that Aboriginal people want, it is anticipated that each ATSIC regional council will provide a forum for the airing of opinions from local communities, and that this will help to empower people even in the smallest communities.

ATSIC is still adjusting to its new role. It is already clear that there are problems with the regional council system, which is under-resourced and has lacked essential support and advice. In 1993 that system was restructured with 36 regional councils replacing the original 60 and allocation of sufficient funding for council chairpersons to be employed full time. The year 1994, when the councillors elected under this new system take up office, will reveal the extent to which these changes have improved the situation. Regardless of these measures the old conflict over the relative priorities to be given to social and economic development is still there, and the new Department, like its predecessors, still has problems in developing programs which effectively link economic and social needs into a more holistic form of development. These issues are clarified in the more detailed consideration of programs which follows.

Recent and contemporary programs for aboriginal economic development in Australia

Self-determination, the process initiated by Australia's 1967 referendum, demanded a radical shift of attitude on the part of many Australians. It also needed practical expression, through programs which demonstrated both to

Table 3.3 Australia, Commonwealth Aboriginal economic development programs

Departments	Programs	Characteristics
Office of Aboriginal Affairs (1968–73)	CCF (1968–73)	Loans for business ventures. Low interest. Mostly individuals
Department of Aboriginal Affairs (1972–1990)	ALC (1973–1980)	Loans for commercially viable businesses. Small scale
	Enterprise funding section (1972–1980)	Support for properties aquired by ALFC
Aboriginal Land Fund Commission (1975–1980)	Land purchase	Land purchase for social and economic reasons (grants)
ALC DAA (ent.) ALFC } Aboriginal Development Commission (1980-1990)	Enterprise program (1980–1990)	Loans for commercially viable projects: grants for others, primarily groups. Land purchase, for social and economic purposes, included
DAA ADC } ATSIC (1990→)	Business funding	Loans for commercially viable enterprises
	Community economic initiatives scheme	Grants for projects of economic and social significance
	Land acquisition and management	Land purchase for social and economic reasons: management likewise
	CDEP	Community employment based on unemployment benefit entitlements
	Regional and community planning	Planning to achieve social and economic development aims

Aboriginal people and to the electorate that the federal government intended to keep its promises. These programs initially included the Commonwealth Capital Fund (CCF); subsequently, after the ALP's return to power in 1972, this program, renamed the Aboriginal Loans Commission (ALC), was complemented by the Aboriginal Land Fund Commission (ALFC) and the enterprise support section of DAA. While these programs were available to all Aboriginal Australians some, particularly ALFC, were especially significant for more remote regions (Table 3.3).

The Aboriginal Loans Commission/Commonwealth Capital Fund (ALC/CCF)

The CCF, which loaned a total of $6.63 million for 450 projects during the five years of its existence from 1968, was specifically directed towards the support of successful Aboriginal business enterprises, either through low-interest loans or by providing security to enable applicants to obtain bank loans. At regional levels, where the Office of Aboriginal Affairs did not have a presence, major trading banks helped in processing and advising on applications. Assistance to primary industry (including pastoral properties) and commercial activities (including retail stores) was important and viability was assessed by an advisory committee under the Office of Aboriginal Affairs. As Palmer (1988: 11–12) comments, the CCF was assimilationist rather than pro-self-determination. It did not overtly support land rights: most loans were for individuals or very small groups rather than communities, and commercial viability was the main criterion for success. CCF annual reports (Capital Fund, 1969–74) clearly show that these problems were recognised. Other problems, highlighted by failure of projects supported by CCF, included lack of Aboriginal management expertise and lack of equity, particularly for rural borrowers who had to face the severe and unpredictable market fluctuations common in primary production.

The prime change following the transfer of the CCF to the ALC in 1974 was the addition of a housing and personal loans component to the fund. Major trading banks continued to be involved, although perhaps to a lesser extent because DAA, under whose auspices ALC functioned, decentralised its operations from Canberra to regions; and primary industry and commercial ventures continued to receive most of the funding.

The Aboriginal Land Fund Commission

The ALFC, established in 1975 as a statutory body to purchase land for Aboriginal groups on the basis of social and economic need, reflected demands for practical recognition of Aboriginal land rights (Palmer, 1988). By the time it was wound up in 1980 its 5 commissioners, 3 of whom were Aboriginal, had made 59 purchases throughout Australia at a total cost of over $6 million. The bulk of the purchases had been in New South Wales, Western Australia and the Northern Territory and, when purchases are considered by area of land acquired, the Northern Territory and Western Australia, where all purchases of large cattle stations were located, were the most important regions (Palmer, 1988: 150–1). More than 50 per cent of purchases in southern and eastern Australia were less than 1 sq. km in area. Annual expenditure fluctuated widely because of uncertainties over funding; the impossibility of predicting what land would come on the market and what applications would come in from Aboriginal groups; and the need for DAA, through its enterprise fund, to provide support for the maintenance and development of new acquisitions. Other difficulties, including its operation during a turbulent period in Aboriginal affairs, and opposition from powerful private-sector groups such as pastoralists' associations and mining companies, also affected it. These problems undermined ALFC's activity and made it less effective than either it or the Aborigines whom it attempted to serve had hoped.

The ALP initially promised to allocate $5 million per year for the next ten years for the purchase of land for Aboriginal groups. However this figure was drastically revised even before the ALFC was established and, following the change of government in 1975 and subsequent cost-cutting measures, their funding became miserly. Over the five years of their existence they received only $6,067,500. Expenditure of funds was further hampered by the restriction of ALFC's role to the purchase of land alone. In the case of pastoral station purchase, a major component, this made it impossible to buy stations on the conventional walk-on/walk-out basis because their funds could not be used to buy the stock or any of the moveable improvements such as vehicles. Without these assets it was impossible for the enterprises to fulfil whatever commercial potential they might have. DAA was responsible for allocating funds for these elements from its enterprise vote, but, until the final year of ALFC's existence, insisted on negotiating a separate agreement for each purchase. They were therefore clearly able to influence the success or failure of ALFC's operations and in some cases properties were lost because vendors became tired of waiting. Although applications from Aborigines far exceeded ALFC's funds, they were not always able to spend all the money available to them.

These problems partly reflect the unstable nature of the fledgling portfolio of Aboriginal affairs during the 1970s; they also stem from confusion over the purpose of the land fund. An Aboriginal land fund's role, implied in the submission of the Council for Aboriginal Affairs in 1971, spelled out in McMahon's Australia Day speech of 1972, and incorporated into the charter of the ALFC when it was established, was to purchase land for social and economic purposes. The commissioners certainly saw it this way and Aboriginal valuation of land, stressing its spiritual and social worth, figured highly in many of their assessments of applications. Properties purchased

109

were not only largely in areas of cultural importance to Aborigines; many were also in marginal grazing country, and degradation and misuse of the land was common. As a result their commercial viability was extremely dubious, and they have never been able to meet the criteria of economic viability preferred by many government people. This issue remains a bone of contention.

ALFC's operations also conflicted with private interests, primarily through the pastoral and mining industries. As Palmer (1988) and Coombs (1978) both record, many pastoral industry leaders were quick to state that Aboriginal pastoralists would undermine the industry and they did not hestitate to use the media to stir up opposition to proposals for the acquisition of land for Aboriginal groups. This could lead to opposition bids, outbuying the ALFC which was forced to operate strictly within the market valuation. Mining companies also became involved. Aborigines from Borroola, in the Gulf country, asked ALFC to buy Bing Bong station for them but Mount Isa Mines, knowing that the area contained significant reserves of silver, lead and zinc, clinched the deal with an offer above the market price (Palmer, 1988: 73–5).

Enterprise section (DAA)

Prior to the establishment of ALFC, DAA's enterprise activities also involved the purchase of pastoral properties, including Daguragu, the scene of the Wattie Creek walk-off of Aboriginal stockmen in 1966 and, in many ways, the catalyst of the contemporary land rights campaign. Subsequently DAA, in line with the general departmental policy of community development, defined as:

> processes by which the efforts of the people themselves are united with those of government authorities to improve the economic, social and cultural conditions of communities, to integrate those communities into the life of the nation, and to enable them to contribute fully to national development.
>
> (DAA, 1975)

took responsibility for support of the infrastructure of all properties acquired. Their enterprise fund was also used to support projects other than pastoral stations, but the overall enterprise allocation from DAA's budget was extremely low (only 4.2 per cent of total expenditure in 1979/80). This was in line with the department's continuing emphasis on its welfare role.

Actual enterprise expenditure varied markedly, not only from year to year but also between states and territories. Not surprisingly the Northern Territory, largely under direct Commonwealth responsibility until it was granted self-government in 1978 and also the region with the highest proportion of Aboriginal population, was the most favoured region and until

1979/80 it accounted for well over 40 per cent of the enterprise vote. Western Australia, also with a high proportion of Aborigines, especially in the northern regions, was the other main recipient with more than 20 per cent from 1976 onwards. These figures reflect the concentration of ALFC's purchases.

The roles of ALC, ALFC and DAA (Enterprise) were all different but at the same time interdependent. Co-ordination of their activities was a major problem. For example, an Aboriginal group living on a pastoral property purchased through the Aboriginal Land Fund Commisssion then had to apply through the DAA Enterprise section for assistance in maintaining and running the enterprise, and might also apply through the Aboriginal Loans Commission for support for other services such as a station store. An additional concern was that, in reality, non-Aboriginal people strongly controlled these bodies, a characteristic which clearly conflicted with the ideals of self-determination and self-management. The establishment of the Aboriginal Development Commission (ADC) was designed to deal with these problems.

Aboriginal Development Commission (ADC)

As stated earlier, ADC saw its role in promoting self-sufficiency as an essential element of self-determination and aimed eventually to achieve this without government funding. This was to occur through the setting up of a capital fund, more correctly called the Aboriginal Entitlement Capital Account. Those plans did not eventuate and all of ADC's programs were therefore paid for out of the General Fund, most of which came from annual appropriations granted through the Budget. With a total estimated budget for 1989/90 of almost $120 million it was clear that the goal of independence from government funding was far from being reached.

ADC initially had three functions, a combination of the responsibilities of the ALC, the ALFC and DAA's enterprise section:

1   to acquire land for Aboriginal communities and groups;
2   to lend money to Aborigines for housing and other purposes;
3   to lend and grant money to Aborigines for business purposes.

Others, such as overall responsibility for housing, were later added. Its economic development functions (categories 1 and 3 above) remained highly significant and in the final year of its operations (1989/90) they accounted for 35 per cent of its total expenditure.

ADC inherited many of the problems of its predecessors, including conflicts about social and economic priorities within its role and difficulties of dealing with poverty-stricken clients with limited experience in management. Applicants for assistance from its enterprise fund were offered loans or grants. The former, which were available to both individuals and groups, required equity of at least 5 per cent, average interest repayments of 5 per cent and were preferred for commercially viable projects. The maximum loan level was $75,000. Many of the more experienced individual applicants, the ones most likely to be offered loans, came from urban and metropolitan communities in southern and eastern states. Grants, in contrast, were more important for people in remote areas; 70 per cent of ADC's grants in the first year of its operation went to the Northern Territory and Western Australia. They were available only for groups/communities and could be offered either for enterprises of dubious commercial viability, or for projects primarily of social significance. Assessment criteria for grants ranged from conventional measurements of economic viability, including profitability and efficient management, to elements such as the cohesiveness of the group; social benefits; and economic benefits flowing on from the goods and services which the project would provide. In effect ADC soon abandoned attempts to distinguish between non-viable enterprises and social ventures.

ADCs enterprise funding, like that of DAA and ALC, went mostly to primary industry and commercial ventures, with grants generally more important in the primary sector and more loans going to commercial activities. In every year grants for the purchase and support of pastoral stations accounted for at least half of the money. As indicated above, many were indeed of dubious commercial worth and the steady decline in the proportion of primary activity grants through time suggests that ADC were increasingly concerned about this drain on their resources. The vagaries of primary industry, particularly in the pastoral sector in remote Australia, also caused problems. Even experienced non-Aboriginal pastoralists in these regions must expect non-profitability in some years, and must rely on capital to survive until better times. Retailing, especially for the construction of new store buildings in remote Aboriginal communities, accounted for a high proportion of commercial funding, both loans and grants.

Difficulties in establishing its role and administering its funds resulted in a recommendation from the Committee of Enquiry into Government Expenditure (1984) that ADC develop a comprehensive corporate plan to cover the next five years of its operations. From 1986 the organisation came under its new planning structure, within which enterprise support remained an important component. The corporate plan stressed that policies concerning enterprises would:

> provide loans, guarantees or grants for increase of capital rather than for recurrent costs in enterprises, through improving access to other funding agencies (e.g. banks) wherever possible; ... give higher priority to loans as compared to grants; ... carry out full appraisal of projects, particularly of the management training and advisory services which might be needed to achieve commercial success.
>
> (ADC, 1990)

As these policies indicate, ADC was increasingly committed to commercial viability as the first priority for enterprise funding; its interest in the support of social objectives declined. Amounts spent on this sector rose from $13.2 million in 1985/6 (18 per cent of ADC's expenditure)to $37 million (over 30 per cent of the expenditure) in 1989/90. Enterprise programs were divided into three main categories: loans to enable Aborigines to acquire or develop commercially viable businesses, or grants to Aboriginal communities for similar projects (the Economic Independence category); grants to enable Aboriginal communities to purchase land or develop social facilities (the Social Advancement category); and, from 1986/7 onwards, grants for management and training advisory services (the Enterprise Support service). Economic Independence received over 70 per cent of funds. In the Economic Independence sector loans were increasingly favoured and by 1989/90 they accounted for over 50 per cent of this category's funds; in 1984/5 only 20 per cent of this component had gone on loans.

ADC's Social Advancement sector, consisting entirely of grants, included land purchase. As with the ALFC, the importance of this varied widely from one year to another, and for very similar reasons. In 1988/9 land purchase for traditional purposes alone accounted for over 20 per cent of the entire enterprise expenditure but three years previously no such land was acquired. Unavoidable delays in completing transactions, especially because of ADC's increasingly bureaucratic procedures; and higher prices putting properties beyond the means of ADC both help to explain these problems. All along, ADC still appeared to recognise the importance of buying land for traditional (social) as well as commercial reasons. At the same time land purchase for non-traditional reasons, primarily the purchase of property in urban areas, increased. During the entire period of ADC's existence its land purchase sector accounted for 20 per cent of the whole enterprise program.

ADC's enterprise support operations were better co-ordinated than those preceding it. However commercial viability has increasingly been perceived as the sole measure of economic development. In the process community development was downgraded. This responsibility remained largely with DAA. Additional inputs from other Commonwealth Departments, notably the Department of Education, Employment and Training (DEET), became increasingly important, particularly after the government's launch of its new Aboriginal Employment Development program (AEDP) in 1987. This program, a direct outcome of an extensive review of Aboriginal employment needs (Miller, 1985), has functioned primarily within DEET's sphere. Because it deals with employment and training, it is extremely important in economic development. Unfortunately ADC, DAA and DEET did not always manage to co-ordinate well, a particular frustration for people living in remote areas. An Aboriginal cattle station group, for example, might have approached ADC for enterprise funds for the project, DAA for funds to improve infrastructure such as fences, bores or power supplies, and DEET for funds to organise a training program and pay trainees. These requests were rarely well co-ordinated, and the results were often chaotic, both for the economic and social well-being of the community.

Conflict and lack of co-ordination also occurred within ADC, between head office in Canberra and regional and state offices, where field staff were employed; and between ADC's sectors themselves. Competition for funding appeared to be a factor in the latter case. Thus buying land for traditional purposes cut into money which might be used for enterprise, and the persistent problem of providing better housing for Aborigines tended to overshadow other demands. The breadth of ADC's role also placed a heavy demand on the skills of its staff, many of whom were having to make decisions without the relevant training in financial matters, or practical experience in projects such as pastoralism or retailing.

Finally ADC, as a large spender of public monies, was continually subjected to scrutiny from other Commonwealth Departments and offices. In 1988 a number of cases of misspending and poor financial monitoring in the enterprise sector led to a review of both ADC and DAA by the Auditor-General. This report (Auditor-General, 1989) commented that 'The enterprise program has occasionally been treated as a welfare program rather than a development program.' This raises an interesting question on the interpretation of 'development' – here it is clearly understood to be synonymous with success in the cash economy; social advancement is excluded. Altogether this report stressed the need for financial accountability and commented on deficiencies in administration, particularly in the approval procedures for grants and loans but also in control over the repayment of loans. Many ADC officers were concerned that the Auditor-General was interpreting the role of their organisation too narrowly, solely within a non-Aboriginal paradigm. They were, however, forced to accept some of the recommendations, including higher levels of control and accountability in enterprise funding. This could be seen as undermining the true meaning of Aboriginal self-management. The Auditor-General's committee, however, did not accept this. They stated that if stricter control 'reduced the high failure rate of enterprises it would be more acceptable in the long run than the demoralising effect of the failure rate' (Auditor-General, 1989: 61). In other words, those communities forced along the commercial road to economic development would be achieving a higher degree of self-determination.

Aboriginal and Torres Strait Islander Commission (ATSIC)

ATSIC's programs specifically concerned with economic development include Business Funding, the Community Economic Initiatives Scheme (CEIS), Land Acquisition and Land Management, Community Development Employment Projects (CDEP), and related programs concerned with

Regional and Community Planning. The ATSIC Development Corporation (ATSICDC), a separate statutory body set up under ATSIC's Act, is also primarily concerned with economic development.

The Business Funding Scheme (BFS) and CEIS together continue the role of ADC's economic independence element of its enterprise scheme. However, while business funding is specifically aimed at lending money to commercially viable projects, CEIS supports community self-determination through grants for income-generating businesses which may also have social and cultural benefits. These principles therefore continue the ADC's allocation of loans and grants on the basis of economic viability. In its initial year ATSIC's BFS disbursed loans totalling over $7 million to 74 clients; a further 82 applicants received grants of over $10 million, some of them for the maintenance and upgrading of infrastructure on pastoral properties considered to have the prospect of being economically viable. Other important funding areas were retailing (store construction) and construction. As with ADC, ATSIC's enterprise funding went disproportionately to Western Australia and the Northern Territory, which together received almost 60 per cent of the allocation; Queensland was the other main recipient. This confirms the continuing importance of ATSIC's economic development program to people in more remote parts of Australia (ATSIC, 1991).

CEIS is a pilot program which deals primarily with ATSIC's additional funding as part of recommendation 311 of the Royal Commission into Aboriginal Deaths in Custody (RCIADIC) that:

ATSIC ensure that in the administration of its enterprise program a clear distinction is drawn between those projects that are supported according to criteria of commercial viability and those that are supported according to social development or social service satisfaction criteria.

(RCIADIC, 1991)

CEIS is concerned primarily with the second type of project, i.e., projects not acceptable to BFS. Funding from 1992/3 to 1997/8 will amount to almost $47 million. It offers grants to communities, often building on enterprises started under CDEP and then found to have a potential to generate income. CEIS therefore links closely with CDEP and also with ATSIC's community planning programs; in addition it has links to two programs in other federal departments which are concerned with land management issues.

ATSIC's Land Acquisition program has until now continued to operate in ways very similar to both the ALFC and ADC programs. A 1992 review of the program revealed that by the end of the 1990/1 financial year land acquisition under all these programs spanning the last two decades had cost, by 1990/1 values, more than $72 million (ATSIC, 1992a: 85). On a state/territory basis Western Australia and the Northern Territory, with 23 per cent of funding each, had been the biggest beneficiaries. Queensland and New South Wales, with 18.7 per cent and 15.2 per cent of funding respectively were the next most important; Tasmania had received only 1.8 per cent of the total. Most Aboriginal groups surveyed in the course of that review said that the acquisition of land had certainly helped them in social terms. Employment, since the widespread introduction of CDEP, had also significantly increased. However few groups had been able to generate a substantial private income. As this suggests, the program has apparently gone far in meeting its social objectives, often the ones most highly valued by Aborigines, but the ideals of generating economic self-sufficiency through land purchase have not been realised. As the reviewers pointed out, this should make ATSIC question whether this program, currently part of ATSIC's commercial section, is located under the correct part of the portfolio. This issue came to a head in 1992 because, following the recommendations of the Royal Commission Into Aboriginal Deaths In Custody (RCIADIC), ATSIC was allocated an additional $60 million over the next five years for the purchase and management support of land. This effectively increases the funding for this program by 150 per cent. Discussions on the policy and program frameworks for land acquisition and management for 1993 to 1997 centred on how to achieve a balance between expenditure on new purchases and the funds required to support and develop those already acquired. In 1993, as an interim measure, the program was absorbed into BFS (Land Acquisition) and CEIS (Land Management). This separation of the two complementary sections of this vital program was viewed with concern by many people including Aborigines, ATSIC officers and others. Additional cause for alarm was the placement of Land Acquisition under BFS because people felt that there was bound to be little sympathy for the applications submitted by groups wanting land to be purchased primarily for social and cultural reasons. Subsequently Land Acquisition and Management were again brought together under BFS but these arrangements are under review. Changes in the structure of the new Aboriginal National Land Fund, announced in the May 1994 budget with an allocation of $1.46 billion over ten years have still to be finalised. Negotiations between ATSIC, other government departments and key Aboriginal organisations over the control of the fund are still in process.

ATSIC also administers the CDEP program, which now accounts for 35 per cent of its total funding. A high proportion of CDEP funds (64 per cent) are effectively additional to ATSIC's general budget, because the Department of Social Security (DSS) is responsible for handing on the unemployment benefit equivalents for all Aboriginal communities on the scheme. By June 1991 ATSIC was looking after 168 CDEP communities, employing 18,000 people in a wide range of part-time jobs. CDEP is, through this highly significant employment contribution, a very important segment of aboriginal economic development. If communities so wish, CDEP can be used to employ stockmen on a cattle property, or people working in the arts and crafts industry, or a gang constructing houses, or office staff. It can also be used to pay people who, because of their knowledge and experience, spend a large part of their time instructing others in cultural traditions, or running ceremonies. It has enhanced the economic prospects of many projects which otherwise would have folded because the precarious state of their profit margins would not enable them to pay their workforce. In remote areas, where these problems are common, it has been particularly valuable.

ATSIC's planning programs, occurring at both regional and community levels, should also fill an important gap in the process of achieving a higher level of self-determination. As yet the success of these programs is uncertain. It remains to be seen whether the doubts expressed by many Aboriginal communities, who have perceived the whole exercise as a waste of time and money and have felt that there has been insufficient consultation at the grassroots level, are justified. Wolfe's (1993a and 1993b) reviews of key issues certainly question whether either the aims of this program or its implementation are really based on self-determination principles.

ABORIGINAL DEVELOPMENT POLICIES AND PROGRAMS IN CANADA AND AUSTRALIA: SOME COMMON THEMES AND ISSUES

Despite some marked contrasts in the earlier circumstances surrounding the development of aboriginal affairs policies in Canada and Australia, the recent policies and programs followed by their federal governments have shown some remarkable parallels. On a large scale these have included: their foundation on assimilationist, welfare-oriented policies and the challenges which these have faced through the transition to aboriginal self-determination; conflicting definitions of development, particularly ambivalent attitudes towards the relative importance of social and economic development aims; and, with their common emphasis on primary resource development as the economic base, an inherent vulnerability which affects the availability of all government funding and puts programs at risk. The co-ordination of aboriginal development support, both between federal government departments and between sections of the same department, is another common theme. Overcoming obstacles such as these poses considerable challenges many aspects of which, despite a number of useful investigations by government and non-government organisations (see, for example, (Australia) House of Representatives Standing Committee of Aboriginal Affairs, 1988 and 1989; (Canada) CARC, 1988), are yet to be met.

DIAND and its Australian equivalent, DAA, have both operated from a welfare-based platform, within which aboriginal development has been perceived primarily as a process which fitted into the broader parameters of Canadian and Australian society and which assumed that aboriginal people would not be capable of taking prime initiatives for either social or economic change affecting them. The welfare approach sought to alleviate the obvious effects of poverty and disadvantage; and any development opportunities which it offered were set within the assimilationist mould. There was little scope for aboriginal Australians or Canadians to publicly voice their opinions either on what alternatives they might prefer or on ways in which these might be set up. Self-determination, the new policy from the 1960s, was theoretically supposed to solve these problems. This still has not happened. Aboriginal affairs policies in both countries demonstrate the co-existence of both assimilation and self-determination. When one realises that many of the individuals working in these bureaucracies have been there for the whole of this period, it is not surprising that the entrenched paternalistic views of earlier times still exert a strong influence. When that influence becomes sufficiently strong as to cause a 'back-flip' in the overall policy statements themselves this causes enormous problems. In Canada the 1969 White Paper and the 1985 Nielsen report (see Weaver, 1981; 1986a; 1986b) show that such backward steps are all too possible. Australia too could easily experience such reversals, probably couched in terms of the thrust for large-scale resource development as the answer to the country's economic woes. As the Canadian experiences show, aboriginal opposition to such proposed changes can today be sufficiently powerful for them to be abandoned. Nevertheless they cause confusion and dissent within the aboriginal community and are therefore destructive to the types of progress for which people are striving.

The definitions of development adopted by the Canadian and Australian federal governments have also varied, both from time to time and from department to department. In general they have increasingly recognised that development has social and economic and environmental components and that it should focus on the long-term sustainability of these three elements. But there have been no determined attempts to support development within this holistic framework. The fragmentation of programs dealing with economic and social (community) issues both within and between departments such as DIAND, ISTC and CEIC (Canada) and DAA/ADC/ATSIC and DEET (Australia) has made this impossible. Thus programs designed to support enterprise development operate as if these enterprises were separate from the overall framework of the aboriginal community in which they are situated. Clearly this is not the case – they are affected by a wealth of social, cultural and political elements and they themselves affect these components of community life. Attempts to deal with these linkages by integrating programs dealing with each component have not been highly successful, and all too often departments take the easy way out. Recent moves to place ATSIC's land acquisition program clearly in the economic section of the department apparently reflect the views of some senior bureaucrats that that is where it ought to belong, and where its administration would be most easily dealt with. They do not reflect the views of those who have benefited from the program, who have been unanimous in their statements that the greatest advantage from land acquisition has been the social and cultural stability conferred through the return of the land to their control. Nor do they acknowledge the program's environmental implications. Here ATSIC lacks the necessary technical knowledge and expertise within its own department but could obtain this by calling on appropriate personnel from other federal departments such as Primary Industry. The co-ordination needed for such an operation has often been deficient in the past but there have been recent signs of improvement. Ultimately the only way in which programs could deal more appropriately with this situation would probably be through a radical restructuring of the departments themselves, not a process which they are going to favour.

Another common theme concerns the effects of the 1980s and early 1990s economic pressures on both the Australian and Canadian governments, pressures which have increased the emphasis on cost-cutting, the main-streaming of government departments and government programs and a demand for strict accountability on the expenditure of government monies. These measures are probably inevitable, given the broader circumstances. But they certainly do not encourage self-determination. The provision of special services for aboriginal people is expensive, particularly with so many living in remote parts of these countries; and many other Australians and Canadians find it hard to accept that this comparatively small proportion of the population merits any special treatment. They feel that they should be able to join the mainstream (to assimilate). This has at various times led to proposals to scrap separate portfolios dealing with aboriginal issues; and, within the departments concerned, to rationalise programs. The latter approach, if based on sound information on what programs are intended to do and the best methods for them to do this, is advantageous. However, if it leads to more and more rigid sets of bureaucratic guidelines which create an inappropriate aboriginal development model, it will be a distinct dis-advantage. Financial accountability, an inevitable element of these pressures, also narrows the development framework. As both DIAND and ATSIC currently accept, this has forced them to assess their performances quantita-tively and hence by methods which must discount the importance of social aspects of development.

Another approach is to encourage people into the private sector. Thus DIAND attempts to release itself from its responsibilities for IEDF, or at least encourage people to use IEDF for loan support/collateral; CAED, and its predecessor NEDP, increasingly emphasise the role of NCCs and see these eventually as local native banking organisations/ credit unions; and all lenders oppose the idea of soft low-interest loans in favour of aboriginal clients paying the same borrowing rates as any other applicants. All of these approaches are designed to wean aboriginal economic development into greater self-sufficiency and, where appropriate, enhance their chances of being competitive with the rest of the mainstream.

These broad questions probably cannot be dealt with within the existing structures of federal aboriginal affairs in Canada and Australia. Radical restructuring of the departments involved would probably be needed. Some more recent developments do however suggest a willingness to experiment, to meet some aboriginal needs halfway. Canada's venture, through NEDP and now CAEDS, into the financing of NCCs is an interesting development. It is obviously fraught with difficulties: how do you account for money once it is being administered by someone else? how do you control the use to which such money is put? etc. There is considerable potential for failures. But there are also possibilities for successes; and the attempts made by the government to hold back from controlling the process have certainly been laudable.

What about the particular impact of these government policies and programs on remote area aboriginal development? Here many of the problems tend to loom larger. Communities are closer to the land, more traditional in their social and cultural values and characteristics, more affected by physical isolation and certainly less well-equipped to join the mainstream. In other words, their ability to share in many of the development opportunities being offered under the Canadian and Australian federal aboriginal affairs policies is quite limited. It is all the more important for them to gain the right kind of support, in the holistic framework which accords with the ideas of sustainable development. The remainder of this book uses four of the main forms of development affecting remote Australia and Canada – land-based development, mining, tourism and retailing – to examine this contention.

4

DEVELOPMENT AND LAND-BASED ENTERPRISE
Living on the land

Today aboriginal people have regained legally recognised title to large areas of country which has made them major land-holders in remote parts of Canada and Australia. Land-based economic activity is particularly important to them – it not only provides them with much of their living, but it is also the prime means through which they express their interdependence with the land. Moreover the land rights movements have not only been politically significant but they have also forced aboriginal people to change their lives more rapidly since the 1970s than in any previous period. Yet governments continue to undervalue the contribution made to human support by these land-based activities. This is partly because much of it is still largely traditional in nature.

Much of the territory now held by Australian and Canadian aboriginal groups is inhospitable. Rugged and fragmented ranges, extensive floodplains and huge areas of elongated dunefields inhibit land communications. Climatic disadvantage, including extremes of aridity and temperature and a monsoon rainfall regime which is less predictable than that commonly occurring in other parts of the tropics is coupled with impoverished soils. And intrusions of feral animals and overuse of natural flora and fauna has resulted in loss of the biodiversity which provides aboriginal people with their subsistence base. Thus in Australia marine species such as dugong and turtle and many of the terrestrial marsupials of the central desert region have either disappeared or are much less plentiful than in the past; and in Canada Inuit and Indians can no longer rely on important resources such as whales and caribou being readily available. These deficiencies in turn constrain the types of land-based activity which can be carried out and limit their effectiveness for either human subsistence or commercial gain. However these are not the only criteria to use to prove that living on the land is worth the effort. Indigenous people value subsistence also in social and cultural terms, and the mere fact of land ownership makes people extremely happy because they are then free to do everything that they feel is necessary for them to express how important land is to them. No matter how poor their country, they will want to use it for subsistence. Such feelings are obviously

121

strongest for those who now control their own ancestral territories, of which they hold both cultural and ecological knowledge. Even for others who, unable to regain their own lands, have been forced to live on country belonging to someone else, such factors are important. They too would be loath to trade the land for money, or to misuse its resources. Land-based activities, of which the main ones considered here are the production of food through subsistence; hunting and trapping for furs and meat; and pastoralism all demonstrate such attachments.

SUBSISTENCE FOOD-PRODUCING ACTIVITIES

Many aboriginal peoples in both Canada and Australia are still heavily committed to harvesting significant portions of their food directly from the land. As is now more widely understood, this activity is of both economic and cultural importance. This combination of values was clearly present in traditional forms of subsistence, where hunting, fishing and foraging were carried out without influence from modern technology or input from the cash economy. It is still present in contemporary forms of subsistence, even when traditional practices have come to include newer and more effective techniques for harvesting and where money helps to make the subsistence system more productive. This modern integration of cash into the subsistence economy is of particular interest in terms of development. It demonstrates how indigenous people have accepted useful commodities from the non-indigenous world to enhance production through their own knowledge and skills.

This discussion covers the following key issues: the contribution of subsistence food gathering to overall sustenance; the technological equipment needed for modern subsistence, what it costs and how this affects the viability of the system; how indigenous peoples value subsistence, both in economic and in socio-cultural terms; what measures can be introduced to ensure that subsistence, if it is indeed of high value to indigenous peoples, is encouraged; and, more generally, the contribution of subsistence to the effective and responsible usage of these fragile lands in remote areas.

Modern hunting and gathering in Australia

Obtaining food from the land, commonly referred to in Australia as collecting 'bush tucker', is not, as many outsiders assume, a haphazard activity, carried on at any time and in any place whenever the opportunity arises. While people eagerly seize opportunities to hunt whenever these arise, careful planning and organisation more commonly play their parts. The following brief account of a typical expedition, from Australia's central desert region, reveals some of the realities of modern subsistence.

122

Women hunting in the desert (from Young, Fieldnotes, various periods, 1978–84)

It's a winter's Saturday morning in Yuendumu, one of the largest Aboriginal towns on the edge of the Tanami desert in central Australia. After a frosty night the temperature rises swiftly with the sun and women set their sights on going 'hunting'. Family groups, consisting of grandmothers, mothers and children, negotiate for transport that can take them to their favoured places for digging for *witchetty* grubs and goannas and collecting 'bush tomatoes', the fruits of the edible varieties of desert *solanum*. These places are in their ancestral country, more than 100 km to the west, beyond the end of the marked roads and in rough territory hard on car tyres and springs. After securing a vehicle belonging to one of their younger male relatives they are off, laden with heavy iron crowbars made from discarded metal rods from the council garage, billy cans from the shop, blankets, water in jerry cans and food – tea, sugar, bread, flour for damper and perhaps packets of frozen meat. The truck is laden down with a cheerful and noisy crowd, eagerly anticipating both the joys of eating sweet 'bush tucker' and also of getting back to their country, to make sure that it is in good heart.

Along the road they are keenly observant, forcing the driver to stop whenever they spot prey: a goanna running up a tree, or into its underground burrow; or edible fruits. Eventually they arrive, and set up dinner camp under a tree in the dry creekbed, close to a soakage where, with only a little effort, fresh water can be obtained from under the sand. Some ensconce themselves here, pregnant women and those whose young children hinder them from moving long distances in the heat of the sun. The rest set out, spreading across the desert harvesting as they go. Billy cans are filled with fruit, abundant after the recent rains. A clump of *acacia kempeana*, the witchetty bush, offers promise of good harvesting and most of the women burrow under the low spreading branches to dig into the root system. Signs of the grubs' presence – discarded pupae cases in the hard earth – confirm the presence of plenty of food and within an hour more than fifty grubs, some larger than index fingers, are added to the fruit in the billy cans. Meanwhile, children, bored with the tedium and hard physical effort of digging, go to hunt for signs of sand goannas. They soon find these – clear tracks of feet and tail and holes hidden under rotting mulga logs – and tentatively prod around with their crowbars until one sinks rapidly into the hollow burrow below. Frenzied digging follows, and with yells of excitement a goanna about 30 cm long is hauled out by the tail and unceremoniously despatched with a quick blow to the head from an iron crowbar. Enthusiasm is renewed, and within another hour half a dozen reptiles meet their doom.

By now it is afternoon. The catch, minus a few fruits and berries eaten en route, is brought back to the dinner camp fire, prepared, cooked and eaten along with tea and damper. Everyone is tired, happy and replete, many commenting that no matter how much beef you have to eat it is only 'bush tucker' that can really fill you up. They linger round the fire talking about the country, and particularly about the significance of catching witchetty grubs, the ancestral beings of this place and the spiritual ancestors of all the women present. Older women sing some of the songs of the land, and answer the children's questions about both spiritual and environmental lore. Eventually, as the sun sinks in the late afternoon, they stir themselves, collect firewood to take back to camp, load up the truck and go. They reach home after dark, and

123

scramble out at their family camps where their menfolk are sitting at the fire, waiting for dinner. 'What about tucker?' the men say. 'What tucker?' say the women. 'We're full of bush tucker and we're tired. We'll just go to bed.'

This brief cameo of subsistence life highlights many features of modern day hunting and gathering - the complexity of its role in Aboriginal society, combining traditional knowledge and beliefs with modern technology and aspirations, satisfying both economic and social needs; and demonstrating above all the strength of the attachment of the people to their land. Accounts from other communities would reveal similar features, although differences in detail, such as the effects of seasonality on the subsistence activities of people in the monsoonal areas of Arnhem Land (Altman, 1987), would be clearly apparent. These broad characteristics of Aboriginal subsistence, combining as they do both economic and social considerations in a cultural setting very different from that of non-Aboriginal society, pinpoint some fundamental differences in attitudes and behaviour between modern Aboriginal hunter-gatherers and those government authorities ostensibly charged with providing practical support for the ideals of self-management and meaningful development.

The subsistence contribution to survival

The contribution of subsistence to Aboriginal sustenance as the major non-monetary component in their economy is, for many groups in remote Australia, of prime importance. It is not, for them, a component that has to be measured. However quantification of subsistence contributions does become an issue if people want to argue for its support as part of development policy. Lack of reliable data and the inappropriate nature of conventional methods of measurement hamper attempts to gauge how valuable subsistence is to Aborigines.

One problem is the very nature of subsistence: the seasonality of the different activities involved; the fact that most of these are gender specific; the fact that the subsistence resource base varies environmentally and hence that findings from one region are not applicable to others; the fact that subsistence has significant cultural value, essentially an unquantifiable attribute; and the fact that subsistence is often combined with other activities, of both cash and non-cash nature. Assessment methods used generally include either imputed values, whereby subsistence products are equated with store-purchased foods according to their costs and type (for example kangaroo = venison; goanna = chicken; fish = fish); and nutritional contribution, whereby the analysis of subsistence foods (for example in terms of energy, protein, mineral content) is examined to determine the overall contribution to the diet. Either method requires assessment of how much food has been obtained.

Detailed Australian evidence on the contribution of subsistence is drawn from a very small number of studies, notably Meehan (1982), Altman (1987) and Devitt (1988). Two of these studies, Meehan and Altman, discuss contemporary subsistence in the Northern Territory's Top End – the tropical wetlands of Arnhem Land – while Devitt's study provides the only comprehensive data on contemporary desert subsistence, and even its usefulness is limited because it concentrates primarily on the activities of women and gives little information on the contributions made by men. All three studies accept the need to measure the value of subsistence if it is to be recognised as a viable economic activity. However all also warn against the dangers of accepting findings unquestioningly. They are particularly wary about the use of imputed values, which inevitably ignore the cultural importance of these foods. Thus, in the above example of desert women's gathering activities, counting the number of witchetty grubs obtained would provide a false impression of the value of the day's work. The fact that they, as women for whom witchetty grubs were of great cultural importance, had also visited that particular country and had been able to demonstrate its significance to their children provided considerable additional value. Meehan, Altman and Devitt also warn that any form of measurement artificially separates subsistence from its social and cultural context. These reservations should temper any conclusions on the size of the subsistence contribution, no matter whether data on imputed values or amount of game/ produce are used.

Accepting the above, studies conducted in Australia reveal that for many Aboriginal people in remote communities subsistence activities still provide significant components of their food supplies. They also contribute raw materials vital for production of tools and artefacts for everyday use, ceremonial occasions and, increasingly, for sale to the outside market economy. Examples of the latter include ochre for artefact, ground and body painting, bark and timber for making spears, boomerangs, and carrying dishes and feathers for decoration.

Studies of the subsistence contribution to Aboriginal sustenance in Australia show that, particularly for people living in small remote communities, it is still considerable. However it is affected by environmental differences, particularly seasonal variations in rainfall, and by the history of land use, which in many parts of the country has severely depleted both natural flora and fauna. Thus coastal areas of northern Australia, a region rich in marine resources and little affected by pastoralism, are richer than the desert.

Altman's study of Momega outstation near Maningrida (Altman, 1987) showed that on average 47 per cent of energy consumption and over 80 per cent of protein came from the subsistence component in the diet. If subsistence production was quantified by imputing values based on the cash

125

Plate 4.1 Setting out for the hunt in central Australia's Tanami desert

Plate 4.2 Controlled burning of spinifex for vegetation regeneration and small game hunting. Aboriginal use of fire is a traditional land management practice which effectively counters the effects of disastrous wildfires

Plate 4.3 Warlpiri women with harvest of *solanum*. These desert fruits are a very good source of vitamin C

Plate 4.4 Witchetty grubs, harvested from the roots of desert *acacia kempeana*

Plate 4.5 Central desert lizards commonly obtained for 'bush tucker': from top to bottom – sand monitor; spiny-tailed monitor; blue tongue lizard

Plate 4.6 Successful young hunters

required to obtain similar foods from retail stores, then the subsistence contribution to incomes exceeded 60 per cent. Seasonal variations showed that the dry season contribution of subsistence was greater than that in the wet, but that it remained a highly significant component of sustenance at all times. Meehan's study at another Maningrida outstation, the coastal community of Kopanga, complements that of Altman because it also takes marine resources into account. She found that marine resources, largely shellfish and fish, contributed between 27 per cent and 37 per cent of energy and 52 per cent and 75 per cent of protein at different seasons of the year (Meehan, 1982). The arid interior would never have provided such rich resources as the tropical north. Nor is it possible to discern such regular seasonal production patterns as in the north because rainfall is much less predictable. In addition much of the arid zone has long been used for extensive cattle grazing, and bush tucker resources have been depleted. Despite this, Devitt's study in a small Anmatyerre outstation about 250 km to the northeast of Alice Springs showed that people obtained 38 per cent of their food by weight from the bush; and that bush foods provided a very high proportion of protein – almost 75 per cent of that consumed (Devitt, 1988).

A common theme in all of these studies is that people in small communities, where population numbers have been sufficiently small to prevent resource depletion near where they live, can obtain much higher proportions of their food from subsistence than can their counterparts living in the larger centralised settlements. For example, when observing the contribution of subsistence to the diets of Aborigines living on Doon Doon cattle station in the East Kimberley I saw that they obtained comparatively little food from the land. They themselves attribute this largely to the effects of cattle grazing. In fact their own cattle, occasionally killed for their own consumption, were now the main form of subsistence their land produces. One could almost say that, for them, cattle had become 'bush tucker'! People living in neighbouring Warmun community similarly made only limited use of subsistence resources, normally only hunting, foraging or fishing at weekends. In large desert communities such as Yuendumu, where most people go hunting and foraging only at weekends and have to travel long distances to reach rich areas, it is likely that only between 5 per cent and 10 per cent of food comes directly from the land's natural resources.

The fact that contemporary subsistence activities, like those of the past, are also gender related further inhibits measurement of their overall contribution. Although many hunting and gathering groups include both men and women each trip often emphasises the interests of one or other group. Thus if a vehicle goes out kangaroo hunting the occupants are primarily men, and any women who accompany them will have little opportunity to walk around seeking goannas to dig for. Conversely if women decide to go to an area particularly rich in witchetty grubs or goannas their husbands, looking for larger game, may well be quite unsuccessful. Results from studies of

Plate 4.7 Cooking kangaroo in the Tanami

subsistence productivity have inevitably been affected by the gender of the researchers. Thus Altman's study probably underestimates women's contributions, while Devitt's specifically set out to examine women's subsistence, and the contributions of men received less attention. Devitt's study is particularly interesting because she collected her data as a participant in hunting trips. This should provide much more reliable information than that obtained through questioning people after their return, because women consume much of their catch before they reach home. Her discussion highlights for the first time the important part played by goannas and lizards, mostly obtained by women, in providing the protein element in desert diets. Current gender differences in subsistence production are also affected by differential access to the new technology, particularly vehicles. On the whole fewer women can drive or are car owners and therefore they are less likely to be able to go foraging when they wish, or where they wish. Even when vehicles are specifically designated as 'women's vehicles' their use cannot always be controlled because reciprocal obligations continually force them to lend them to others, often to male kin.

While all these examples demonstrate the importance of subsistence in the remote communities studied, it must be stressed that not all communities would produce the same results and that there might well be significant changes over time. Variations in the subsistence component depend on environmental differences; differences in type of community and its access to resources; and differences in the knowledge and skill of those involved

129

(younger aboriginal people, born and raised in the centralised settlements, may well be less highly skilled hunters and gatherers than their parents). Subsistence production also varies seasonally. Altman's (1987) study of Momega shows that almost 60 per cent of sustenance came from subsistence in the late dry season months, but only 40 per cent in the middle of the wet season. Finally, the relative significance of subsistence to total community incomes is affected by the value of other inputs. If cash incomes rise, perhaps through an increase in jobs or in wages, subsistence may appear to be less important. This does not necessarily mean that people are less active; on the contrary, more cash may well allow them to obtain more food from the land. However the subsistence share in the total would then be smaller.

Traditional and modern technology in subsistence

As the above discussion shows modern subsistence is now inextricably linked to the cash economy. Although Aboriginal technology still includes traditional items such as wooden coolamons, fish spears and harpoons many of these tools have been improved by use of stronger materials, particularly metal and nylon. Aboriginal women today use metal crowbars rather than wooden digging sticks, and traditional techniques for catching large fish species such as barramundi have largely been abandoned in favour of nylon lines, lures and nets. Rifles have almost entirely replaced spears and boomerangs as tools for obtaining larger game such as kangaroos. Harpoons are armed with metal barbs attatched to the shaft with nylon ropes. Most importantly vehicles and modern boats with outboard motors are now vital necessities.

The costs of obtaining equipment for hunting and gathering, particularly vehicles, are important factors for Aborigines. However, because subsistence is an activity which appears to bring little cash profit, funding agencies have been reluctant to accept that vehicles are needed. In the late 1970s and early 1980s the Aboriginal Benefit Trust Account (ABTA), a slush fund derived from royalties from mining activities occurring on Aboriginal lands in the Northern Territory, was extensively used to provide four wheel drive vehicles for remote communities. Between July 1978 and December 1984, 431 such vehicles were purchased, mostly for outstations (Altman and Taylor, 1987: 29–31). In ABTA eyes, the prime purpose of these vehicles was to give people a lifeline, a communication link to the larger Aboriginal communities where most of the essential services were located. However outstation people wanted them for hunting, and used them heavily for this purpose, often in rough country. The wear and tear, which often resulted in the vehicle becoming obsolete in a remarkably short space of time, was seen by many government people as evidence of wrongful use. ABTA subsequently clamped down on applications for outstation vehicles and in the next year only $1 million, sufficient to buy fifty vehicles, was allocated for this

Plate 4.8 Women fishing for estuarine barramundi on Australia's Gulf coast

purpose. Considering that ABTA funds come from royalties and hence could be seen as part of the compensation paid by mining companies for their use of Aboriginal land, restricting the purpose of applications in this way seems somewhat ironic. At that time Aborigines saw little problem with the fund being nicknamed the Aboriginal Benefit Toyota Account; but the government did.

Support for subsistence food production

Living off the land obviously, from evidence such as this, provides aboriginal people not only with vital social and spiritual support but also, in more remote communities at least, adds significantly to their total incomes. Few people understand this fully. Yet this contribution decreases the extent of the Aboriginal socio-economic disadvantage. Income disparities, such as those shown in census data, are for some groups at least, probably smaller than they appear to be. This is particularly the case for more isolated communities. Fisk's (1985) comparative analysis of the real incomes of different types of Australian Aboriginal community showed that in the most remote communities, the small outstations, people's combined incomes from monetary and non-monetary sources might in fact be greater than those of people in the larger centralised settlements. Altman's detailed study, which provides the basis for part of Fisk's analysis, showed that the per capita cash incomes of people in Maningrida, the centralised settlement, exceeded those in the associated outstations by 44 per cent; however, when the subsistence component was added to cash earnings outstation people had incomes which were 75 per cent higher than their counterparts in Maningrida (Altman, 1982).

Figures such as these speak for themselves and can surely be used to encourage funding bodies to support Aboriginal subsistence. However that is not the only reason why such use of land and resources is important in Australia. This is arguably a more sustainable way of using marginal lands in remote areas than introduced forms of land use such as pastoralism. In arid Australia particularly it is becoming increasingly clear that reduction of cattle numbers on Aboriginal-owned cattle properties has improved the chances of regenerating the native vegetation which provides vegetable foods and wildlife habitats, and as a consequence some of the ravages of soil erosion have decreased (Young et al., 1991).

Recognising the contribution of subsistence is only one part of the problem. Even when the value of subsistence is acknowledged the contemporary nature of the activity is imperfectly understood. In earlier times in Australia Aboriginal people were encouraged, and sometimes forced, to leave the mission settlements or cattle stations during periods when their labour was not required to go 'walkabout' and 'live off the land'. That practice may have been quite acceptable to Aboriginal groups eager to get back to their country and escape from the restrictions of station life. It also saved money – decreasing the government's contributions for rations and the need to pay workers a full-time living wage. This attitude, that subsistence should make no demands on the cash economy, still to a large extent prevails. But present day food gathering, to be efficient, requires modern technology and hence requires cash. Support for present day subsistence activities must realistically be based on that assumption.

As yet no comprehensive schemes for subsistence support have been introduced for Australian Aborigines. Cash benefits from their subsistence, which generally does not involve hunting and trapping for furs and skins for sale in the market economy, appear small although, as Altman and Taylor (1987) discuss, such an assumption ignores the vital subsistence input into the production of goods for the lucrative arts and crafts market. However Altman and Taylor (1987) further discuss the possibilities of introducing an Aboriginal subsistence support scheme for people living in outstations. Their proposed Guaranteed Minimum Income for Outstations (GMIO) would provide people who are determined to stay in these small remote communities with enough financial support to encourage them to persevere as hunter-gatherers. Related production, such as arts and crafts manufacture, would also be encouraged. GMIO, unlike unemployment benefits, would not require work tests and would be more flexible than the existing systems. It would essentially be a form of income support. The challenge of introducing GMIO, or alternative forms of income support, has yet to be taken up.

Hunting and trapping in Canada

Aboriginal people in remote Canadian communities also place high value on subsistence activity and depend heavily on both meat and vegetable products from the land for their sustenance. However, in contrast to the situation in Australia, hunting and gathering for 'country food' in Canada has been inextricably linked with the trading of part of the catch, mainly furs and pelts, into the non-aboriginal market system. Indeed, as summarised earlier, hunting and trapping has arguably been the prime instrument of colonialism in Canada, with organisations such as the Hudson's Bay or Northwest Companies deliberately tying aboriginal people into the capitalist economy from the earliest stages of their contact history. This interlinkage of hunting and trapping for the market and hunting and foraging for country food makes it impossible to realistically assess the contributions of these two elements separately. The following account from the northern boreal forest region of Northwest Territories clearly demonstrates the totality of that integration.

Summer and fall on the Mackenzie River (Smith, 1986; Young, Fieldnotes, 1984)

By early July the great Mackenzie River has completely thawed at Fort Good Hope, a small town with a population of about 450 Hare Indians around 500 km south of the delta on the shores of the Arctic Ocean. It's high summer, school children are on holiday and potentially making nuisances of themselves in town, and the mosquitoes and blackflies, always worse in the main settlement, are present in huge swarms. Family groups decide to move out to fish camps, travelling in small boats along the river to places where there is fresh drinking water, good places to set nets and a breeze to blow the mosquitoes away. The cores of these groups are small, perhaps only a handful of adults and children, but while they set up temporary residence in camp for several weeks, other members of the family frequently visit for a day or two or longer.

Along with their boats, their equipment consists of other forms of transport such as three and four wheel motor scooters, useful for quick visits to town, nylon nets and sets of metal hooks, set in favoured locations and visited twice daily to check on the catch. Fish are not the only product. Rabbits are snared near the riverbanks behind the camp and waterfowl obtained whenever they appear. By August, with the end of the short three month summer approaching, the birds begin to return in large numbers to the river as they migrate southwards from the high Arctic and considerable numbers are shot for food. Away from the river moose, bears and, in the early autumn, migrating caribou are hunted whenever signs of their presence have been spotted and when enough men are around to arrange an expedition. Other smaller game, beavers and muskrats, might also be taken but this would depend on the current value

Plate 4.9 Summer camp on the Mackenzie River near Fort Good Hope; skins drying
in the background, and being cleaned in the foreground
Source: Photograph by Shirleen Smith

of their pelts and the availability of other game.

Food and pelt processing is very important. Elderly women in the camp direct the cutting and drying of the fish, to be stored both for winter food for their families and for the dogs. If the family still has a dog team the quantities required will be large. Moose meat is dried and the hides are tanned for making moccasins and mukluks. Other skins, perhaps wolf or bear, are cured and stored, either for sale to the Fort Good Hope Hudson's Bay Store or to the local HTA Co-operative, or for making winter clothes needed for the long periods spent in the winter trapping camps.

By late August the seasons are changing. Temperatures drop, mosquitoes disappear and sleet storms herald approaching winter. The people return to the comfort of their solid log cabins in Fort Good Hope, where they delight in meeting up once more with friends whose summer fish camps have been in other distant locations and begin to make plans for the winter's trapping.

In late September, when snow begins to accumulate and the rivers have already started to freeze over, it is time to leave again. About thirty families, almost one-quarter of Fort Good Hope's population, depart to spend most of the winter at their traplines, mostly away from the great river near small creeks and lakes in the interior where, for generations, they have held rights to trap animals. Their HTA assists in the complex organisation of winter camping – arranging air transport if needed, advancing cash to allow people to buy enough provisions for several months, and to purchase guns, ammunition, vehicles, fuel, tents, snowmobiles and whatever other equipment is needed, and ensuring that everyone has the means to communicate with base so that safety in the rigours of the northern winter is as assured as possible. This time it is the skilled

134

hunters and their wives, the able-bodied adults, who leave. Older people and school age children remain behind. By the end of October the bustle of departure is over and Fort Good Hope, now blanketed in snow and ice, settles down to the less frenetic life of the long winter nights.

The contribution of hunting and trapping to the economy in northern Canada

In general attempts to assess the economic contribution of subsistence have been both more detailed and more numerous in Canada than in Australia. This reflects the obvious incorporation of subsistence into the monetary economy, explicitly through trapping for furs and pelts. Harvesting surveys, often conducted by government departments and aimed specifically at assessing both types and quantity of animals being obtained by trappers, and the substitution of imputed values of country foods have been the most common assessment methods used. Problems arising with these approaches have been similar to those highlighted by researchers in Australia. As Usher and Wenzel (1987) discuss, official harvesting surveys usually depend on hunter recall, while special harvesting surveys carried out by individual researchers are specific to particular communities and it would be dangerous to generalise from these results. Usher (1976), like his Australian counterparts, has also warned against the unquestioning use of imputed values to assess country food production because these do not overtly recognise its social and cultural significance. Other Canadian and Alaskan researchers, including Freeman (1976), Muller-Wille (1978), Lonner (1986) and Smith (1986) have also warned against simple acceptance of quantification based on imputed values as a means of reckoning the subsistence contribution to sustenance. As in Australia, therefore, evidence for the subsistence contribution in northern Canada can only be accepted with reservations.

An additional problem, as Usher (1976) and Muller-Wille (1978) have both pointed out, is that it is usually impossible to separate food for human sustenance from the production of all the other items obtained in the course of hunting and trapping. People collecting country foods also obtain a whole range of products used for both artefact and tool production. These include various timbers and barks used to make containers, items such as porcupine quills or feathers used in clothing manufacture, and, for the Inuit, soapstone for the production of carved figures. Animal skins are of major importance. Food and all of these products are interdependent. Fish, for example, is obtained to feed dogs, which provide the transport needed to visit traplines and take pelts to the trading post; but some of the fish catch is also consumed by the hunters.

Regardless of these problems, all Canadian studies which have specifically set out to assess the importance of country food have found it to be significant. As anticipated, environmental and historical land-use factors affected the level of that contribution and access to the rich resources of marine and riverine environments obviously enhances its value. In addition, as in Australia, the level of involvement in country food/hunting production appears to vary between different types of community, with the proportion of hunters in smaller, more remote communities generally being higher than those in larger communities or among groups living in smaller and larger towns (Ames *et al.*, 1989: Ch. 1).

Quigley and McBride (1987), discussing the traditional sector in the economy of the Hudson's Bay Inuit community of Sanikiluaq in 1984/5, showed that, with imputed food values, country foods accounted for an annual income of $7,079 per capita, 57 per cent of community income. Cash income from the sale of furs was by that time negligible, largely because of the effects of the embargo on sealskin sales introduced in 1983. Seals had formed over 75 per cent of the catch for food in that community. Quigley and McBride estimate that if the sealskin embargo could be overcome (perhaps by processing skins for leather rather than fur) the cash income from furs and skins would add about 25 per cent to the present community cash income. Smith and Wright's (1989) study of the contemporary economic status of hunters in the Inuit village of Holman in Northwest Territories confirms the importance of the country food contribution in particular. They showed that full-time hunters, the prime producers of both country foods and pelts for sale, on average obtained country foods which, from imputed values, accounted for food to the value of over $17,000, at a profit of almost $15,500 when harvesting costs were deducted. This exceeded their actual amounts of cash received from selling pelts and furs, guiding game hunters and taking part in other related wage earning activities ($13,021 per hunter). In all cases, since they obtained more country food than they consumed (enough to feed another four people per year), they also helped to sustain other members of the community who, as full-time wage earners, lacked the time to spend on subsistence. As Smith and Wright (1989) point out, the importance of these contributions is clear. But government reluctance to provide overt support for full-time hunting as a way of life is a cause for concern (see below).

Canadian studies conducted amongst Indian communities in the boreal forest environments have also confirmed the importance of the subsistence contribution. Rushforth (1977), working with the Dene people of Beaver Lake in the lower Mackenzie basin estimated that in the early 1970s between 27 per cent and 43 per cent of food was coming from subsistence. Feit (1982a) and Salisbury (1986) produced higher estimates for the Cree, who they said obtained at least 50 per cent of their food from the bush.

In general the Canadian studies, whether conducted before or after the destruction of the sealskin markets, show that the income from the sales of furs and pelts from hunting and trapping is always well below the costs of production in terms of capital equipment and other needs. However when country food production is added the picture is very different – it is worth so much that the activity becomes economically viable. The total incomes of full-time hunting families then equal or possibly even exceed the levels reached by full-time wage earners in remote communities.

Hunting and trapping also, as indigenous people frequently stress, has other non-economic values. Taking part in these activities welds the relationships between people and land and enables the cultural identification with animals, birds and fish which forms the basis of traditional spirituality. Learning to participate in trapping is an experience essential to the preservation of Canadian aboriginal identity. Thus:

> Hunting and trapping represents much more than simply a source of money income.... We also respect the animals we capture in order to ensure the benevolence of the lord of the animals, and their reproduction in generations to come. In addition, we transmit to our children knowledge of the animals' habits.
>
> (Attikamek-Montagnais Council, 1986; quoted in House of Commons, 1986: 11)

As Usher (1976) stresses, the contribution of subsistence itself should not therefore be measured as a separate entity, but as part of the total lifestyle into which it is incorporated. This characteristic of subsistence does not readily accord with the program guidelines imposed by external agencies charged with providing monetary support. This, as further discussed below, hinders people from devising alternative forms of development.

Cash inputs into hunting and trapping

As Usher (1972), Muller-Wille (1978) and Freeman (1982) describe, Inuit and Indian groups now depend heavily on snowmobiles for land transport, on rifles for hunting and on engine-powered canoes for reaching their fishing grounds. Aircraft are even coming to play a part in modern subsistence, as a means to reach hunting grounds in inaccessible country. As described above, organisations such as the Fort Good Hope HTA in the lower reaches of the Mackenzie River outlay considerable sums of money to hire planes to transport their members to hunting camps at the beginning of the winter season, before lakes and rivers are sufficiently hard frozen to allow overland transport (Young, 1991a).

All of today's subsistence activities therefore require money. As Smith and Wright (1989) have shown Inuit hunters had to spend considerable amounts of cash to pay for vehicle fuel, for store food to last them until they had obtained good supplies of country foods, and for general maintenance of equipment. They suggest that in Holman such costs were worthwhile because of the profits obtained ($17,117 worth of food at a cost of $1,669 per hunter). However it is easy to imagine that in areas where food supplies are less assured, but where people are still extremely keen to carry on subsistence activities, the returns would be much smaller. In the season of 1984 most of the $12,000 outcamp grant allocated by the Government of Northwest Territories (GWNT) to the Fort Good Hope people went on the plane costs and it is doubtful whether production would have covered these.

A common criticism of hunters who insist that they need cash to cover their equipment costs is that, if their ancestors managed with dogs, then they should also be able to do so. Such criticisms generally fail to comment that the dogs also had to be fed, presumably on large quantities of country foods. Today modern transport and technology would be the only way of obtaining this food efficiently. Smith (1986) comments on how the decline in the use of dog teams cut down the additional amounts of fish required by Dene hunters. Stories of the advantage of having a means of transport which, in times of extreme deprivation, can also be eaten, have their point. However, because of the centralisation of aboriginal settlements and their growing populations and because people now want to own motor vehicles and snowmobiles it is very unlikely that people will opt to return to the dog-team days.

If people cannot obtain external funds to meet the costs of modern hunting and trapping they obviously have to provide these for themselves. This means earning wages, either through selling artefacts or furs and pelts, or by joining the wage workforce. A number of Canadian examples highlight that, while people in the workforce obviously have less time to spend on hunting than do their full-time hunting counterparts, wages are often used to pay the costs of subsistence. In this way, as Hobart (1981; 1984) has shown, the wages earned from introduced activities such as work on mining camps can help to support the traditional economy. Relatively few aboriginal people have been employed by mining companies working in remote locations. They have lacked the required skills and/or have been perceived to be unreliable workers. However those who have been employed 'on rotation', i.e. where they can alternate between working in the mining camp and staying in their home community for short periods, not only suffered less social stress from living away from home in an alien environment, but also had time for subsistence. Hobart's study of Coppermine, an Inuit community affected by the operations of petroleum companies in the 1970s shows that, from the figures available, returns from hunting apparently did not decline even when people were wage earning; and he concluded that access to cash had actually made subsistence production more dependable because people were able to afford to buy snowmobiles and could visit the best areas whenever they wanted to. His analysis is based only on fur harvesting statistics and therefore refers specifically to the hunting and trapping component of subsistence. However it can be assumed that food production, which complements these activities, would also be maintained. These findings suggest that, contrary to common expectations, industrial development need not always have a detrimental effect on remote aboriginal communities. The cash it provides may be used to support other economic activities, including subsistence production.

Another important issue is the link between welfare payments and hunting and trapping as a way of life. With the chronically low levels of wage employment in remote communities many families inevitably depend heavily on welfare payments for their cash incomes. It has been suggested (House of Commons, 1986: 16) that not only has the trapping industry never been viable, but that it should be discouraged because it hinders aboriginal people from making a determined effort to join the mainstream economy. Such a view, as aboriginal people have strongly voiced, totally disregards the all-encompassing role of hunting and trapping in social as well as economic terms. In fact many welfare recipients, like their counterparts who are fortunate enough to hold wage jobs, use their cash resources to enhance their involvement in hunting and trapping. As this shows, such land-based activity means much more to them than the accumulation of material wealth alone.

Supporting subsistence hunting and trapping as a way of life

Estimates of the total country food harvest for Canada's Northwest Territories in 1988 were in the order of $55 million per year (Ames et al., 1989).This surely is reason enough to ensure that, in economic terms alone, the productivity of that harvest continues. Other arguments for supporting the industry concern the contribution of fur trapping, especially in social and cultural terms. Arguments such as these have led to a number of proposals for the provision of income support for Canada's indigenous hunters and trappers, both to ensure their vital economic contribution but also, above all, to support hunting as a 'way of life'. Schemes include programs such as the Income Security Program (ISP) introduced for the James Bay Cree (La Rusic et al., 1979; Feit, 1982b; Salisbury, 1986) and proposals stemming from much more recent studies conducted on alternative resource strategies for native populations in Northwest Territories (Ames et al., 1989; Usher and Weihs, 1989).

ISP, introduced as a component of the compensation package which was part of the James Bay and Northern Quebec Agreement, is designed to assist Cree hunters to make sufficient income, from both monetary and non-monetary sources, to make it worth their while 'staying on the land'. Its introduction reflected not only recognition of the value of subsistence foods in the Cree economy, but also the fact that the cash returns gained from hunting and trapping are not sufficient to cover the costs of outfitting the activity. As Salisbury (1986) summarises, ISP appears to have been successful in encouraging Cree subsistence. It has resulted in an increase in the number of hunters, a decrease in dependency on welfare and general regeneration of the whole community economy. There have also been social and cultural spin-offs, such as the maintenance of hunting skills, including their transmission from older to younger Cree. Problems have included the complex procedures used in calculating each participating hunter's payment package, and the rigidity of some of the rules which applicants have had to follow.

Usher and Weihs (1989) discuss a Wildlife Harvesting Support Program (WHSP) as one component of a strategy to support the internal economy of NWT. This is further elaborated by Ames *et al.* (1989). As they show, existing government support for NWT subsistence harvesters in 1987–8 averaged approximately $6.7 million, well below what the estimated 5,500 subsistence harvesters in the region would need to meet the costs involved in carrying out their activities. They suggested a WHSP costing between $10 million and $30 million per year depending on the number of participants, and funded either through the land claim settlements under negotiation (then Nunavut and Dene/Metis) or directly by government departments such as DIAND, or CEIC. As they point out, these figures were only estimates because of lack of appropriate data on hunting populations and their harvests. More accurate data would be needed for proper costing.

Obstacles to the trapping industry

In the late 1970s/early 1980s, films of the annual commercial slaughter of Canadian seal pups caused outcries against cruelty to animals. Complaints were made by many individuals and organisations, including animal rights and animal welfare groups. They took full advantage of global media attention and in 1983 an embargo on the culling of seal pups was imposed. The complete collapse of the European and North American sealskin markets followed. In the latter part of the 1980s this was followed by threats of strict embargoes on the sale of all furs. As many aboriginal and non-aboriginal commentators have stressed, (for example, Keith and Saunders, 1989; Wenzel, 1991), this promises a dire future for Inuit and Indian people in northern Canada. Trapping is the only way they can augment their meagre cash incomes. Confrontations, emphasising the rights of humans as well as animals and highlighting the dearth of non-aboriginal knowledge about the relationships, both economic and social, between people and animals in the Canadian north, have been numerous and bitter.

PASTORALISM

Remote parts of Australia, unlike the Canadian Arctic, were attractive to the non-Aboriginal settlers because they offered opportunities for commercial pastoralism. The Australian outback is climatically harsh and unpredictable and is extremely isolated from main centres of population in the southern

Plate 4.10 Droving cattle in Australia's rangelands

states. As a result pastoral operations have to be extensive, with each station requiring large tracts of land to support comparatively few stock. Pastoralism has therefore led to widespread alienation of land from its customary owners, the Aborigines. At the same time, from the beginning, Aborigines were needed as the labour force for the industry. Their environmental knowledge and skills not only proved useful in themselves but they also helped Aborigines to master the techniques necessary to become good stock-workers. As many people now clearly acknowledge, they formed the essential backbone of an industry which has become synonymous with Australian outback life (Cowlishaw, 1983; Berndt and Berndt, 1987).

It is scarcely surprising that in more recent times, as a number of outback pastoral leases have come under Aboriginal ownership, cattle stations have been seen to offer unique opportunities for development in remote areas. It has been presumed that they will provide people with the means to reach economic self-sufficiency. Subsequent events suggest otherwise. Many Aboriginal properties have failed to achieve even a modicum of commercial success. Examination of the reasons for these failures suggests that non-Aboriginal assumptions on their potential profit-making worth were often poorly founded. As with subsistence activities, Aboriginal pastoralism is not purely a commercial one. It also operates within the fundamental social and cultural contexts which reflect the relationships between people and the land.

These, along with other complex historical, environmental, economic and political factors all play their parts in determining how pastoralism can contribute to Aboriginal development.

Historical aspects

The alienation of Aboriginal land for pastoralism inevitably caused conflict. It resulted in losses on both sides, with Aborigines commonly spearing cattle and, under pressure, killing settlers; and non-Aborigines retaliating by killing Aborigines, or at least chasing them away from their country. In some remote areas such as Central Australia and the Kimberley such confrontations occurred within living memory and strong feelings of fear and loss still persist (Ross, 1989; Ross and Bray, 1989). But while such experiences have given many Aboriginal groups very negative feelings towards pastoralists, others, as McGrath (1987) has documented, have from the beginning worked in tandem with their non-Aboriginal employers, and have taken great pride in their work and in their contribution to the industry. Of crucial importance has been the extent to which Aborigines have been able to remain on their ancestral country after it has been alienated for pastoral use. Direct confrontation often resulted in the complete dispersion of the original Aboriginal group and their replacement by other Aborigines who were not customary land owners. Spiritual aspects of land ownership might then be disregarded. But when Aborigines became the stockworkers on their own territory they were also able to maintain their customary spiritual responsibilities.

Such differences are still important today because they continue to affect the operation of contemporary Aboriginal-owned pastoral properties. In general, those Aboriginal groups which remained on their ancestral lands as stockworkers have had greater social autonomy than those who have moved elsewhere. This has given them a strength which has helped them both to make plans for the future and to carry these plans out, even when this has caused conflict with the ideas put forward by external non-Aboriginal agencies. Those groups which lack close spiritual identification with the land within their property boundaries, are less socially coherent, more likely to find that the community as a whole does not support their plans, and hence they are less able to cope with outside pressures.

Two neighbouring central Australian cattle stations exemplify these differences. Willowra/Mt Barkly Aboriginal communities, with a long-standing, stable and mutually respecting relationship with the former owner, exhibited a social coherence and communal determination to work together which assisted them, at least in the early years, in making efforts to assume control and in making plans for the future of the station. Ti Tree people, in contrast, demonstrated greater signs of community stress, and also had difficult relationships with non-Aborigines. This was possibly due to previous antagonism from a repressive pastoralist who discouraged Aborigines without jobs on the station from living on their traditional land. This had curbed their ceremonial activities. On Mt Allan strong mutual respect between Aborigines and non-Aborigines existed but, because the previous owner had left no opportunity for Aboriginal decision-making, the new Aboriginal owners were reluctant to take on that role. Since the former owner had become the new Mt Allan manager it was doubly difficult for the Aborigines to break out of that earlier subordinate relationship, and to recognise that they, as the employers rather than the employees, theoretically were now in control (Young, 1988b).

The history of government policies and programs for Aboriginal economic development is another important factor. The assimilation era emphasised the training of Aborigines for conventional wage jobs which would encourage them to accept non-Aboriginal work ethics and help to absorb them into mainstream society. Cattle projects, introduced during the 1950s on some of the large government and mission settlements, such as Hermannsburg, Haasts Bluff and Yuendumu in central Australia, reflect the operation of this policy. Aborigines trained through these projects were expected to provide a seasonal labour force for neighbouring cattlemen. The pastoralists did well from the deal. Their Aboriginal stockmen did not then qualify for award wages, and they were not obliged to provide good living quarters or other facilities for them. This type of situation was not restricted only to the past. Until quite recently the state-owned properties now held by the Woorabinda people in Queensland were worked for the financial benefit of Queensland's Aboriginal Welfare Fund and the people themselves derived little if any direct benefit from them (Dale, 1992).

Such practices disadvantaged Aborigines in a number of ways. Their wages did not properly reward them for their labour; they often had to live in conditions of extreme squalor; and on both the pastoral properties and in the government and mission sponsored cattle projects they received no training in any aspect of management. The highest rank attained was that of head stockman. All of these factors have important implications. They have had to learn about pastoral management from scratch, including the practical day-to-day technical skills of bore and fence maintenance, the logistics of marketing stock through the abbatoirs and the complexities of financial matters. These problems have undermined the commercial viability of their properties and made it difficult for them to compete successfully with their fellow non-Aboriginal Australian pastoralists.

Environmental aspects

The contemporary operation of Aboriginal pastoral enterprises is significantly affected by environmental deficiencies of many of their properties. This is a result of the limited funding available for the purchase of leases and of decades of misuse of rangelands through over-grazing.

The purchase of land for Aboriginal groups dates primarily from the period following the election of the Whitlam Labor government in 1972. The purchase of two stations already under protracted negotiation – Pantijan (Western Australia) and Willowra (Northern Territory) – was immediately completed as an early expression of the ALP's commitment to land rights and self-determination, and subsequently over fifty stations have been bought throughout Australia. Funds have come principally from the Commonwealth government, under the auspices of the Aboriginal Land Fund Commission, the Aboriginal Development Commission and now the Aboriginal and Torres Strait Islander Commission. However these funds have always been limited, and as a consequence most properties purchased have been cheap, often because they have been badly run-down and have become financially devalued. Many are either on the extreme margins of viable grazing land or were heavily and seriously degraded when Aboriginal ownership commenced. Two of the East Kimberley Aboriginal properties – Doon Doon and Bow River – were badly degraded at the time of purchase; their fences and water bores were in disrepair; and there was no accurate assessment of either the size or composition of the herd. Similarly, in the Northern Territory, Ti Tree station to the north of Alice Springs was described as follows when under investigation for government purchase in 1976:

> The cattle ... would give the impression that the herd is only an average to poor [one] by Central Australian standards.... The main trucking yard is just not useable.... The homestead yard is not worth having. The station is in default of its covenant, requiring two more bores to be sunk.
>
> (DAA 77/166)

Chilla Well (Corranderk), a remote Northern Territory property purchased in that year, had effectively never operated as anything more than a holding paddock, used for watering stock in transit from the East Kimberleys to Alice Springs across the Tanami Desert. Similar stories could be repeated for many other Aboriginal cattle stations. These circumstances have obviously exerted a strong influence on commercial performances. Somewhat surprisingly these inherited deficiencies have received little comment from government officials responsible for funding the cattle enterprises, and as a result the Aborigines are usually unfairly blamed for shortfalls and failures in commercial performance.

The overall extent of environmental degradation on Aboriginal owned pastoral land is unclear. However recorded observations of the condition of these properties at the time of their purchase suggest that there must be some severe problems. Specific studies of the condition of Aboriginal pastoral land have been conducted in some regions but results are inconclusive. In the

Kimberleys, Resource Condition Surveys carried out by the WA Depart-ment of Agriculture in 1990/1 showed that only two of the eight Aboriginal properties had large areas of land in poor condition. This result was better than expected. However other defects did receive comment. These included locally severe erosion at pressure points around soakages, bores, tanks and ponds; the need to introduce properly conceived and controlled management practices, such as regular burning of vegetation to reduce the risk of wildfires and enhance pasture growth; and the need to control feral donkeys and brumbies (wild horses). Aboriginal recollections of the state of the land in the recent past contradict these findings. They say that the land has not only been eroded, but has also lost many species of animals and reptiles and the kinds of vegetation common today are not as useful for supporting either humans or animals. Ledgar (1986) suggested that these feelings should be a good enough basis for the government to provide funds to attempt regeneration programs on Aboriginal pastoral land. This proposal has not yet received support. However, more recently, this question has been taken up by the major Aboriginal Land Councils in the Kimberley and the Northern Territory. Pilot studies for their land assessment project have been supported for 1993 in Central Australia. This will cover all types of Aboriginal land, including pastoral lands, and will provide Aboriginal pastoralists with much better information on land capability. The end result should be much more efficient and sustainable management techniques. Recent attempts to work out realistic strategies for Aboriginal pastoral enterprises, based on a combination of environmental, economic and social elements, also demon-strate enhanced understanding of the situation (Stafford-Smith et al., 1994).

Understanding the environmental limitations is crucially important for considering development and Aboriginal pastoral enterprises. Land degrada-tion, including various types of soil erosion, destruction of biodiversity and the introduction of feral animals and noxious weeds, have had severe effects on the Australian environment, particularly in more marginal country. Over-grazing by hard-hooved stock such as cattle and sheep have destroyed the natural vegetation cover and exposed land to large-scale water and wind erosion. Many of the more nutritious and palatable plants have disappeared, to be replaced with plants such as spinifex grass which are only a valuable fodder when immature, and as a result the grazing capacity of many rangelands has been sadly reduced. Moreover feral animals such as brumbies and donkeys have proliferated on the pastoral lands, where they compete with the cattle for scarce foodstuffs and also contribute to vegetation loss and soil erosion. In 1983 one of the Aboriginal properties in the East Kimberley was reputedly carrying ten donkeys to every head of cattle. Eradication of feral animals in such rugged country poses huge practical problems, which are compounded by Aboriginal reluctance to destroy animals for which, culturally, they feel a strong sympathy (Young et al., 1991). Recently donkey culling programs have been much more successful. Altogether, between 1978 and 1991, over 400,000 donkeys have been slaughtered throughout the Kimberley region (North Kimberley Land Conservation District, 1991).

Social aspects

The association between traditional land responsibilities and contemporary land management on Aboriginal pastoral stations is usually very strong. This has both advantages and disadvantages. Advantages, in Aboriginal terms, come from the social and economic integration of the enterprise. Thus in Aurukun in Cape York the mission established a cattle enterprise which, instead of being centralised, was run in small units on a dispersed basis. Each unit had a group of stockworkers caring for cattle in their own traditional country. This gave people an opportunity not only to enhance their skills in pastoral work, but also to maintain their knowledge of that land for which they were spiritually responsible (Dale, 1992). On a number of Central Australian cattle stations (Willowra, Mt Allen, Ti Tree) a high proportion of the company directors and stockworkers are also traditional land owners, and this seems to create a better basis for decision-making and planning than would otherwise occur. The Ngarliyikirlangu cattle company at Yuendumu, although theoretically an enterprise belonging to the whole large community, has had a similar arrangement. It draws its

Plate 4.11 Cutting out cattle for branding in the yard

workers and directors largely from traditional owners of the country where the cattle graze, rather than from other families whose country lies elsewhere. Other positive effects from incorporating traditional land ownership with responsibility for the contemporary pastoral enterprise include making the best use of detailed environmental knowledge, minimising arguments and enhancing co-operation.

Non-Aboriginal people have, however, seen the incorporation of traditional land ownership into pastoral management much less positively. Traditional owners are not necessarily the most knowledgeable or experienced as far as pastoral operations are concerned and do not therefore always make the best managers. Their mistakes could be costly. On the other hand it may be difficult to run the enterprise effectively without them. On cattle stations where traditional association with the land is much more tenuous, for example on Doon Doon station in the East Kimberley, it has certainly been harder for people to decide how to run the enterprise. Here the constitution of the Aboriginal company wrongly specified that the Woolah people, thought to be the traditional owners of the country, would be the directors; later it became apparent that Woolah country only covered part of Doon Doon and that other land interests were involved. However non-Woolah people were still excluded from holding responsible positions in the enterprise management, and as a result they were reluctant to participate. This inhibited the growth of the community, scarcely viable with only 30 to 35 people in 1986. It also excluded skilled and experienced individuals from making positive contributions.

Economic aspects

Discussions on development and Aboriginal pastoral properties focus above all on their economic situation. Success, measured in terms of the generation of profits, the creation of wage employment for the resident community, and improvements in capital infrastructure and in investment incomes, is what most people expect any Aboriginal pastoral company to aim for (Young, 1988a; 1988d). The results, in those terms, have been very disappointing. The evidence for the poor economic viability of Aboriginal pastoral properties – lack of improvements, poor herd structures, undercapitalisation, lack of managerial and financial experience – is overwhelming. Counteracting such massive disadvantages means a significant injection of funds. Government agencies (DAA and ADC) have recognised this by providing stations with recurrent funding for salary and wage payments, stock improvement, fencing and improvements to water supplies and many other needs. In almost every case comparison of such investment with profits generated, in terms of value for money, has thrown doubt on the value of this strategy. Between 1976 and 1987 Doon Doon, for example, received $535,000 from DAA and ADC but made no more than $120,000 through selling cattle. Evidence such as this has made government agencies increasingly wary of supporting many Aboriginal pastoral activities. Basic questions which must be posed are: firstly, why has funding not been successful in promoting commercial pastoralism? and, secondly, are the current criteria for funding inappropriate in any case?

Reasons why funding has failed to enhance commercial viability include unrealistic assessments of what is possible, and the lack of long-term planning. Many pastoral operations were scarcely viable at purchase, even as a living for one non-Aboriginal family. With some Aboriginal pastoral station communities having populations in excess of 300, commercial viability in terms of an independent economic resource base for the whole group, was never possible. Properties such as Doon Doon were so run down that complex plans for their operation had no chance of success without very substantial investment. Doon Doon, initially valued at $1 million in 1972, was finally bought by the Aboriginal Land Fund Commission for $240,842 in 1976 but by the time the purchase was completed much of the stock had been removed and the lease boundaries had been redistributed to exclude the most valuable pasture on the station.

Planning for the development of that station, and for most others, has been based on the assumption that what is successful for a non-Aboriginal pastoralist will also work for Aborigines. Thus the emphasis has been on large-scale, capital-intensive operations, such as those run by large companies owning a number of properties; labour-intensive operations, obviously appropriate with the large size of the Aboriginal pastoral communities, have often received little consideration. In reality many Aboriginal pastoral operations would probably only succeed commercially if operated on a small specialised scale. And even then their profits would not support the community; they would merely provide a source of independent income, very important in combating the overwhelming welfare dependency of such settlements but not sufficiently large to exclude the need to seek additional family income support elsewhere.

Unrealistic expectations are compounded by lack of long-term planning. Most solutions to problems have come from rapid decisions, made because resources were available at one particular time and no other. They are band-aid solutions. ATSIC, and its predecessor ADC have more recently tried to deal with this by requiring pastoral companies to present a development plan when they applied for funding. However these plans may still be unworkable. Most have been drawn up by consultants working within briefs which are far too narrowly defined.

The second question relates to the criteria for government funding. Funding is based on the idea that a pastoral operation is distinct from the community to which it belongs – an assumption which, given the particular social characteristics of an Aboriginal pastoral station community is patently false. Thus funds are concerned solely with the profit-making component of pastoralism, not with vital elements such as its employment needs and how employment contributes to the community; or to the infrastructure required by members of the community if pastoralism is to flourish. These components are defined as the responsibility of other government departments/funding bodies. Several departments or organisations have to pool their resources if an appropriate funding package is to be created. This co-ordination is remarkably hard to achieve. Development plans for Doon Doon station in the mid-1980s, for example, involved: ADC, for the pastoral enterprise; DEET, for funds to pay for a training officer in pastoral management; and WA housing and essential services departments for long overdue improvements in the community infrastructure. Although all these forms of support were eventually provided they were not necessarily made available at the best time – training money did not come until after the end of the dry season, when mustering had already ceased.

Aboriginal pastoralism and development

Social, environmental and economic factors such as these clearly affect the potential for Aboriginal pastoral enterprises to achieve commercially successful economic development. However when development is considered from the Aboriginal point of view the situation is different. Here, when it is possible to combine the social and cultural interests of the community with the opportunity to generate cash, pastoralism may have much to offer. One key question concerns the types of pastoralism which people want to practise. Aboriginal pastoral activities are very varied. At one end of the scale they consist of 'killer' herds, often maintained by small outstation groups who want to provide themselves with cheap meat; at the other end they are commercial enterprises, run from the homesteads of large cattle stations by methods very similar to those used by their non-Aboriginal neighbours. In between these two extremes lie a wealth of different situations – communities which want to combine a small-scale commercial cattle business with other forms of land use such as subsistence; and pastoral projects which, because they were originally established for training purposes rather than to make a profit, have always lacked the necessary capital investment and infrastructure to become commercially viable. No two of these situations are identical. However all too often government funding agencies have ignored this diversity and have concentrated on developing standard sets of rules and regulations to guide their programs. Those who do not fit the mould do not receive support. Some illustrations of the difficulties involved follow.

Many 'killer' herd pastoral enterprises are associated with outstations on Aboriginal land, for example in parts of the Pitjantjatjara homelands or on former pastoral lease such as Utopia. These are not commercially oriented and only require minimum funding for the maintenance of fences and water supplies in paddocks. With low stocking rates degradation should not be a major issue and the financial management problems which beset larger cattle enterprises do not affect them. Being on extensive areas of Aboriginal-owned land the whole operation remains within both the social and economic control of the community. However 'killer' herds have also been suggested as a suitable form of land use for Aboriginal groups living on small plots of land excised from non-Aboriginal stations. This situation is different. Many excisions are very small in relation to the size of their Aboriginal populations. In the Northern Territory, for example, the average size of excisions granted to date is only 500 hectares, and the average population is 30. Opportunities for running sustainable 'killer herds' under such conditions would be very limited. State Aboriginal Affairs officials in Western Australia have also expressed concern about the effects of overstocking of cattle in the very small excisions granted to Aborigines in the Kimberley. Thus perhaps people on excisions should not be encouraged to keep 'killers' but should instead purchase animals from neighbouring pastoralists whenever they want them – the common practice anyway. While this would overcome the problem of lack of land and overuse of resources it would not necessarily be popular because people have come to view 'killer' herds as a subsistence resource, for which they should not have to pay cash. This is a hard question to resolve. One small community may be allowed to keep a 'killer' herd while the one next door is denied that chance. And if 'killer' herd pastoralism is to be supported, who will fund it? The enterprise section of ATSIC? Or the community development section of ATSIC? In the past government bodies have evaded this question, and 'killer' herd support, which is useful because it helps to provide supplementary meat supplies for a low-income population, has been left largely to Aboriginal organisations such as the Institute for Aboriginal Development and the Central Land Council in Alice Springs.

ATSIC responds in a much more positive way when asked to fund commercial pastoralism. This fits their program guidelines. And, in addition, commercial Aboriginal properties may well be able to gain advice through many of the support services also available to non-Aboriginal pastoralists. Profits from these commercial ventures can help to reduce economic dependency on government funding and services, even if they do not actually provide a living wage for every family in the community. The chances of commercial viability are greatly enhanced if these communities have CDEP schemes which can be used to cover the stockmen's wages. This means that the profits generated by the enterprise can be reinvested, or can be used for other community purposes. However there can be pitfalls with following the wholly commercial path – the community may well lack the necessary management and financial expertise and will then be forced to employ non-Aboriginal managers. This, as Cowlishaw (1983: 55) and others have pointed out, increases the dependency of the people when, ironically, the acquisition of the property was supposed to make them independent. If the non-Aboriginal manager remains in control of day-to-day operations, or the non-Aboriginal government employee or accountant keeps the books without

Plate 4.12 Butchering a 'killer'

discussing finances with the members of the community, such dependency will linger on.

Most Aboriginal pastoral enterprises lie somewhere in between, recognising the importance of the economic independence coming from commercial operations but also benefiting from all the other advantages which flow to them from land ownership. A major problem here is that no funding agency is established to deal with these mixed agendas. Because of this many Aboriginal pastoral communities have been forced into requesting support for a fully commercial operation when that is not what they actually want.

ATSIC is not oblivious to the problems stemming from the misfit between what people want to do with their cattle, and what the government guidelines will allow. The department has tried to find appropriate ways to support Aboriginal cattle managers without taking over too much control from the community. This has included setting up a Management Services unit for a number of Aboriginal cattle stations, as has occurred in West Kimberley; and trying to combine the management tasks of neighbouring properties, so that one person can cope with the whole job. Ultimately the best solution must come from the Aborigines themselves, with the establishment of their own pastoral support agencies. Unfortunately this probably means outside funding, and as yet this has proved a stumbling block. The Central Australian Aboriginal Pastoralists' Association, based in Alice Springs, foundered after

151

only a couple of years. More recent expansion of the land management role in Land Councils, particularly the Central Land Council in Alice Springs, has led to a revival of this idea and ATSIC's recent grant of funds to Kimberley Land Council to conduct a detailed assessment of the present situation and future prospects of Kimberley Aboriginal cattle stations suggests that they are now placing greater emphasis on the provision of Aboriginal-controlled pastoral support services.

DEVELOPMENT AND THE USE OF THE LAND'S RENEWABLE RESOURCES

Renewable resources – the land and its water, soils, flora and fauna – should provide the core for aboriginal development in remote areas. The advantages of enterprise activities based on such resources are that they use locally available commodities and that the skills which they require are widespread within the resident aboriginal population. Activities such as subsistence hunting, fishing and foraging, trapping and pastoralism are means of sustenance which can make huge contributions to aboriginal support, not only in terms of meeting their nutritional needs but also improving their health and well-being. Moreover most activities on the land provide people not only with food or the means to generate cash; they also provide a total fulfilment in people's lives – the opportunity to spend time on the land, to look after it and to identify with it and its resources in cultural and spiritual ways. Activities such as hunting and trapping are not conducted in isolation from the communities to which people belong but are effectively the essence of these communities. As Canadian writers such as Freeman (1985a and 1986) and Usher (1986 and (with Bankes) 1986) have frequently stressed, the use of these resources by aboriginal people demonstrates the total integration of ideas of property rights and all elements of human support. These land-based activities therefore provide the most appropriate means to development and accord most closely with the model of sustainable development discussed earlier.

On the whole Australian and Canadian government agencies have still failed to grasp the significance of this. Despite programs to 'keep people on the land' the general perception that cash aspects of activities such as hunting and trapping or pastoralism can be separated from their non-cash components persists. In general the non-cash component has been ignored. This is partly because it is harder to quantify, and therefore assess the value of this element. It is also because assessment methods reflect government preoccupation with conventional economic forms of development. The challenge of developing methods for assessing both the total contribution of subsistence and other land-based activities is one which needs to be taken up. This is not only important for funding agencies. It is also important for aboriginal groups which, with their rapidly growing populations, must make the best possible use of the land under their control. Appropriate assessment of land capability would have to include not only conventional components such as soil fertility and water quality but would also have to acknowledge aboriginal ideas on land use for cultural as well as economic reasons. Aboriginal constructs of land, including such valuation, would be a vital part of this (Young, 1992b).

Knowledge of what the land in remote parts of Canada and Australia can offer for the support of human populations is currently very sketchy. In Canada, where neither agriculture nor commercial grazing are prime options in remote regions, this is hardly surprising. In Australia the lack of information about the rangelands has, until recently, prevented the development of good management practice. Today, with current attention paid to sustainability rather than to quick returns, and with new techniques in land assessment some of these defects can be remedied. One interesting initiative is the recent establishment of a land assessment program, not under government agencies but under an Aboriginal organisation. This program, in its pilot stage in 1993/4, will develop assessment criteria which reflect not only environmental and economic attributes of the land but also Aboriginally defined social and cultural attributes. Geographic information systems make it possible both to analyse these in ways which present a total picture of the land's potential, and to cover the huge areas which lie within Aboriginal lands. Eventually, funding permitting, this land assessment program will be extended from its initial focus on Aboriginal lands in the central desert to other regions, notably Arnhem Land and the Kimberley. The end result should be much better information to enable Aboriginal land owners to plan their present and future land uses; and to enable government agencies to assess what kind of support and how much support is needed.

Another possible extension of renewable resource use for aboriginal people in remote areas concerns the exploitation of introduced resources which are now firmly established. Experiments with reindeer herding in northern Canada (Nasogaluak and Billingsley, 1981; Nasogaluak, 1983) have indicated that there may be a future for such enterprises, and that aboriginal peoples have the interest and can acquire the skills for working in such projects. Recent resurgence of interest in making commercial use of central Australia's wild camel population also suggests that this type of enterprise may offer attractive opportunities to Aborigines, some of whom still remember working with camels in the earlier part of this century. Government interest in such possibilities is now being strongly expressed through agencies such as Australia's Bureau of Resource Sciences, part of the Department of Primary Industry. Amongst other things their program for Aboriginal Rural Resource Initiatives (ARRI) has recently been focusing on projects concerning Aboriginal use of wild animal resources. While these include commercial development of native species use, such as kangaroos, crocodiles and emus, they also extend to feral animals such as pigs, goats, donkeys, water buffalo and camel (Wilson *et al.*, 1992). Prospects such as these are interesting not only because of their potential contribution to Aboriginal sustenance from the land, but also because they may assist in dealing with Australia's significant feral animal problem.

5

MINING – THE PRIME NON-RENEWABLE RESOURCE OF REMOTE REGIONS

The lands of remote parts of Canada and Australia lack the resources necessary for agriculture. But they are rich in other resources, particularly the non-renewable resources sought by the mining industry. Mineral wealth has attracted development to these regions on scales far greater than those affecting any other type of enterprise. In conventional economic terms this has been advantageous. However there have also been disadvantages. Major issues include the impact that mining makes on the fragile Arctic and desert environments, an impact which became globally recognised because of events such as the Alaskan oil spill. Another issue, less commonly publicised, is the impact of mining on the aboriginal residents of these regions. Important topics include their involvement in this form of development, its impact upon them in both positive and negative ways and particularly the extent to which they have been able to use it to further their own development aspirations.

MINING AND THE ABORIGINAL COMMUNITY: SOME GENERAL CHARACTERISTICS

In both Australia and Canada recognition of the mineral wealth of the remote areas came early in the period of non-aboriginal exploration, and was a major catalyst for an influx of population from other regions. Discoveries of valuable metallic minerals in particular, such as gold, silver and copper in the Yukon and Northwest Territories and gold in central and northern Australia, attracted individual prospectors and wealth seekers in legendary rushes which were responsible for the rapid growth of communities such as Dawson City and Yellowknife (Canada) and Halls Creek, Granites and Arltunga (Australia). Unfulfilled prospects and the tolls taken by the rigours of the environment subsequently led to the abandonment of many of these 'towns' and today their presence may only be marked by piles of stone left over after the painstaking grubbing of the miners, rusting equipment and machinery, and the headstones in the abandoned graveyards.

Some, like Yellowknife, survived. There the minerals were sufficiently valuable for larger-scale ventures, operated by mining companies rather than

155

by individual prospectors, to be set up. The profits in these types of operation were large enough to support investment in increasingly complex technology, so much a hallmark of the contemporary mining industry. Large-scale mining came to epitomise the new wealth of remote Australia and Canada and was recognised as the prime way in which these 'barren' regions could contribute to national income. Later the recognition of other valuable mineral resources, including bauxite, manganese, iron ore and, more recently, diamonds in Australia, and uranium and oil and natural gas in both countries, has vastly extended the distribution and impact of large mines. Their effects on aboriginal peoples have been very different from those of the early peripatetic prospectors.

A number of basic characteristics distinguish development based on modern mining activities from the types of development in which aboriginal peoples in remote areas can most easily participate. These include the large scale of mining operations and their capital intensive and highly technical nature. Associated with these is the development of towns, often owned or controlled by the mining companies and with populations consisting largely of people who have migrated into the area from elsewhere and who may well be of different ethnic or cultural origins from the local resident groups. They also include attitudes towards environmental matters, which, because of the emphasis of the mining industry on extraction for greatest commercial benefit, tend to be unsustainable if not openly destructive. In addition the interests of this type of mining development, whether in ownership, control or marketing, lie firmly outside the local region.

The immense scale of modern mining operations, as O'Faircheallaigh (1991: 251) stresses, is the main feature which distinguishes them from other forms of economic activity in remote regions. It is an outcome of the basic desire in the industrialised world to maximise profits by using economies of scale to their fullest extent, particularly in remote areas where associated costs of development, in terms of infrastructure and of the provision of the necessary human resources, are also so high. The result is not only a large mining operation but, in most cases, a town which by local standards is also large. Such towns have very different types of housing and services from those found in neighbouring aboriginal communities. Modern mining towns like Jabiru in Australia's Northern Territory or Norman Wells on the Mackenzie River often look very out of place, southern suburbia transferred without adaptation to the 'bush'. They largely house mining company employees. Many of these people have been associated with the industry for years and have moved frequently between several similar settlements, rarely spending more than a few years in one place. Because of their mobility they have had little opportunity to establish long-lasting commitment to the place where they live. Because of the nature of the industry most of these workers are highly skilled and very experienced in their jobs. For aboriginal people, living in much smaller, more scattered settlements with much more limited infrastructure and facilities, these settlements are obviously attractive. However, mining company towns often have strict rules over who can live there, who can use the services which they provide and even who can visit the town casually. Because of this local aboriginal people may in fact have very little direct contact with these mining towns.

The contrast between the Aboriginal and mining communities in the Argyle area of Australia's East Kimberley region is particularly stark. Here the town, developed and administered by Argyle Diamond Mining Inc. (ADM), is off limits for everyone without an entry permit. Aboriginal people travelling between the small cattle station communities where they live must divert around the town area, and cannot use any of the services available although some of these, for example health and retail services, are very limited in their own settlements. And while the roads of the town of Argyle and its airstrip, large enough to take the jets which connect it directly with the state capital of Perth over 5,000 km away, are wide and fully sealed all other roads in the immediate surrounds are only surfaced with gravel and continually dusty. Over 95 per cent of Argyle employees are non-Aboriginal and, since this mine operates on a fly-in/fly-out basis, are either unmarried or are working away from home, leaving their families in distant Perth. ADM management have tried to provide for the recreational needs of their workers within the town. But the workers, not surprisingly, have felt restricted and some have taken to fishing or swimming in the surrounding country. Here there has been some conflict with local Aboriginal groups, who have been accustomed to using the same areas. Argyle is perhaps an extreme example of the physical separation of the aboriginal and non-aboriginal communities. However even when aboriginal people can freely go into mining towns they still often feel unwelcome and alienated.

Environmentally, modern mining operations have been destructive. The removal of a non-renewable resource usually causes some environmental damage. In remote Australia and Canada the minerals have often deliberately been removed within a very short time and, particularly in the past, less expensive open-cut methods have been used. Whole hills in Australia's Pilbara and Kimberley regions have disappeared during iron ore and diamond mining enterprises. Such destruction, and its secondary effects on water courses, soils and vegetation, demonstrates that this type of mining is not only an unsustainable form of land use in itself but that it also threatens the sustainability of other forms of land use. For aboriginal peoples the effects on native fauna and flora, on which the subsistence component of their economy depends, are of grave concern. While catastrophic events such as the effects of the *Exxon Valdez* oil spill on the wildlife of the Alaskan coast are widely publicised, smaller-scale problems of this type – the destruction of local fish stocks in small creeks near a mine or disruption of caribou migrations – occur more often. These have received less attention and aboriginal protests about these types of problems are less likely to be heard.

157

Mining also destroys places of cultural value. This is particularly important for aboriginal people. The Kimberley Argyle Diamond Mine coincides with a barramundi dreaming site of great spiritual significance and, some fifteen years after mining began, conflicts between traditional Aboriginal owners and the mining company still fester. If mining, or other resource development, can be legally opposed through land rights legislation, as occurred at Coronation Hill in the Northern Territory, these conflicts can escalate to the highest political levels.

The third major contrast between mining development and local interests is that of ownership/control and marketing. Mining companies operating in remote Australia and Canada are dominated by external interests, both national and international, and the major markets for their products are almost always elsewhere. Not surprisingly this fosters development attitudes which take little account of local interests as compared with national and foreign interests, and which in environmental, economic and social terms override the small-scale needs and aspirations of the permanent residents of these regions. Moreover because mining is controlled by outside companies and is affected by fluctuations in external markets it can be a vulnerable

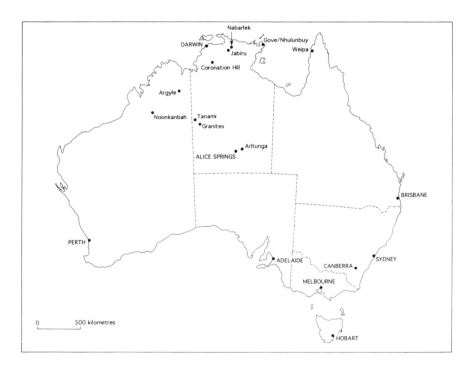

Figure 5.1 Australian mining projects

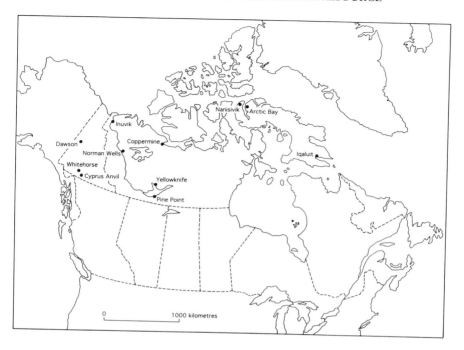

Figure 5.2 Canadian mining projects

enterprise. When mines have to close the effect on local aboriginal employment may be negligible, because so few of them had jobs. However there can be important indirect effects, especially if local groups have held some equity in the project or have been able to use the goods and services provided in the mining town.

IMPACT OF MINING ON REMOTE ABORIGINAL COMMUNITIES

O'Faircheallaigh's (1991) overview considers the interaction between mining and aboriginal communities under three main headings: economic, social and environmental. As he stresses, these do not occur in isolation but affect one another, with, for example, the environmental effects of mining having both economic and social consequences. The following discussion highlights both positive and negative aspects of the economic and social consequences of mining, illustrating points with reference to a number of specific developments. In Australia these include the Argyle Diamond Mine (East Kimberley), the Ranger and Nabarlek uranium mines (Arnhem

159

Plate 5.1 Norman Wells on the Mackenzie River

Land), the North Flinders gold mine (Tanami desert), and the bauxite mines at Nhulunbuy and Weipa (Figure 5.1). In Canada they include oil and gas developments ranging from the shores of the Arctic Ocean to communities far up the Mackenzie River (Figure 5.2), but brief references to lead/zinc mines in NWT and Yukon are also made. All of these mining developments have been subject to some degree of social and economic assessment, both by outsiders and by local aboriginal organisations. In this they are perhaps rather unusual. Many other Australian and Canadian resource developments have taken place without any consideration of the ways in which they have affected aboriginal groups.

The case-studies

Norman Wells

Oil was first struck at Norman Wells, in the central Mackenzie Valley, in 1920, but the development of the field was hampered in its early years by isolation from southern markets. In the early 1940s concern about possible Japanese invasion of Alaska led to the Canol (Canadian Oil) project whereby Norman Wells oil was piped to a refinery at Whitehorse. This caused a sudden tenfold increase in production, but by 1946 this had declined to its pre-war levels. Thereafter production again increased slowly and by 1980 it had once more come close to its 1946 level. By then the significant increases in oil prices of the 1970s, coupled with increasing demand and improved technology, had once more made the Norman Wells field an attractive proposition. The construction of a pipeline to transport oil and gas to Alberta and points south was a key component of this expansion. During the construction period of 1982–4 the large influx of population quadrupled Norman Wells to a town of over 1,200 people (Bone and Mahanic, 1984). This caused considerable anxiety among the local Dene/Metis population. Negotiations carried out between the federal and NWT governments, the mining companies and the Dene/Metis included agreements for social and economic impact monitoring and planning of the project and these studies highlight many issues of relevance to this discussion (Bone, 1983; Bone and Stewart, 1987).

Metallic ore mining in northern Canada

During the latter half of the twentieth century northern Canada saw a number of important ventures involving the mining of metallic ores in very remote areas, often adjacent to existing aboriginal settlements. These include the Pine Point mine to the south of Great Slave Lake, the Cyprus Anvil mine near Whitehorse in the Yukon, the Rankin Inlet nickel mine on the western side of Hudson's Bay and the Nanisivik lead/zinc mine at Strathcona Sound, on the northern end of Baffin Land. As Macpherson (1978a and 1978b) summarises, all initially faced problems concerning the costs of development and transportation of ore to smelters and markets. Their impact on the local aboriginal populations has varied with time. In the earlier developments such as that at Rankin Inlet, there was little attempt to recognise either that there would be a social impact, or that steps should be taken to ensure that aboriginal people gained some economic benefit from the enterprise. Employment opportunities were limited, both because of the nature of the work but also because of lack of access between the mine and the local communities; and there was no provision for royalty payments or compensation monies. More recent mines such as Nanisivik have seen a higher level of involvement with local groups. Pine Point, where production commenced in 1964, represents the middle period. Here the aspirations of aboriginal people were recognised but understanding of how these might be incorporated into company policy was still very limited. In the early 1970s very few people from Fort Resolution, the nearest aboriginal community to the mine, had worked on the project, and in general people felt that Cominco, the mining company involved, had failed to listen to them.

161

Argyle Diamond mine

The Argyle Diamond Mine (ADM), developed by the Ashton Joint Venture Consortium including CRA as a major partner, started production of its rich deposits in a remote area of Western Australia's East Kimberley region in 1987 (Dixon and Dillon, 1990). It is not only the most important diamond mine in Australia but reputedly one of the richest in the world, with at least 15 per cent of its products of gemstone quality. ADM's company town now has a population of around 300 to 400 people, largely shiftworkers operating from Perth on a fly-in/fly-out basis. The source of the diamonds, the kimberlite pipe, coincides with an important Aboriginal spiritual site, representing ancestral barramundi dreaming. That, combined with the lack of communication between the miners and the Aborigines during the exploration phase, negotiation problems over the mining agreement, and the fact that Aborigines in that area are severely socially and economically disadvantaged compared to others, has caused major concern about the whole project. In addition, the lack of land rights legislation in Western Australia means that Aborigines have had little control over the mining operations themselves, and little opportunity to negotiate worthwhile compensation and profit sharing deals. These problems make ADM a particularly useful and telling case-study.

Plate 5.2 Argyle Diamond Mine from the air

Ranger and Nabarlek Alligator Rivers uranium developments

In the 1980s the Alligator Rivers region on the western edge of Arnhem Land in the Northern Territory's 'Top End' proved to be of global importance in uranium production. That was not their only claim to fame. The Ranger and Nabarlek mines, which came into production in 1981 and 1980 respectively, were the first to operate under agreements which conferred substantial royalty payments on Aboriginal people. By 1986 Ranger Uranium Consortium and Queensland Mines Ltd, the two companies involved, had paid over $70 million to the Aboriginal Benefits Trust Account; 70 per cent of these monies were then disbursed to the NT Land Councils and to Royalty Associations representing the traditional land owners, with the remainder forming ABTA's funds used for meeting development needs of Aboriginal communities throughout the Territory. Because of such arrangements these mines obviously have an impact on Aboriginal economic development which is felt well beyond the local area.

Important factors affecting the development of these mines and their interaction with local Aboriginal groups include the fact that they lie on Aboriginal land (Nabarlek), or on an excision surrounded by Aboriginal land (Ranger) with that Aboriginal owned land leased back for administration as Kakadu National Park; that the wealth which they have generated for the local Aboriginal groups has been highly significant; that the two groups involved, the Gagadju and Kunwinjku Associations have, as O'Faircheallaigh (1986b) discusses, used this cash in different ways, with the Gagadju people generally favouring investment with positive long-term implications while their fellow Kunwinjku members have favoured rapid expenditure on consumer items; that, because of a detailed social monitoring study carried out at the existing Aboriginal town of Oenpelli (Australian Institute of Aboriginal Studies, 1984), the effects of this cash on Aboriginal people have been well documented; and that Jabiru, the new town constructed to house the non-Aboriginal workforce and with a population which is now well over 1,500, has not only become an important attraction for Aboriginal visitors but has blossomed into a regional centre with an additional important role, that of providing services for the growing number of tourists who come to Kakadu National Park (Lea and Zehner, 1986). All of these factors make this an extremely important case-study of the interaction between the mining industry and Aborigines.

Gold in the Tanami Desert

The Granites goldmine lies on Aboriginal freehold land in the remote Tanami desert to the northwest of Alice Springs in Australia's Northern Territory. The gold-bearing deposits which it exploits were first recognised early in this century when a number of gold rushes occurred. Nobody became rich and mining was abandoned. North Flinders Mines, the current operating company, began production in 1986 with a mine based on a fly-in/fly-out workforce housed in a town which at any one time would have a resident population of around 300 (Howitt, 1991). The Warlpiri, the local Aboriginal group who are the traditional owners of the land, provide very few mineworkers. Negotiations between North Flinders Mines and the Warlpiri, represented by the Janganpa Association, resulted in on-going payments consisting of both up-front monies, annual payments and additional sums varying with the profits of the mine. Between 1984 and 1989 this had totalled almost $4 million. During that same time Janganpa had received a further $75,000 from the Tanami joint venture, largely in payment for water and exploration rights at another site, about 200 km northwest of the Granites. Discoveries here and at Granites itself suggest that future production in the region as a whole will be much greater than at present.

Weipa and Nhulunbuy bauxite developments

The production of bauxite from Comalco's mine at Weipa, on the west coast of Queensland's Cape York peninsula, has been of major importance in global terms for over thirty years (Howitt, 1992a). However it is a development over which the local Aboriginal people, many of whom were displaced from their former settlements and who congregated on the outskirts of the new mining community, have exerted little control. Dating from the 1960s, when discussions over self-determination were not yet prominent, and being situated in Queensland, where land rights legislation has until very recently been conspicuous by its absence, no royalty-type agreements were signed. Hence there has been no assured economic base for the dispossessed Aboriginal community. Aboriginal employment in Weipa, a town of over 3,500 people, has been low and Aboriginal groups are clearly marginalised. Weipa provides an interesting comparison with the Northern Territory's bauxite mine, developed by Nabalco at Nhulunbuy in northeast Arnhem Land (Howitt, 1992b).

Nabalco's bauxite and alumina venture at Nhulunbuy, in northeastern Arnhem Land, has been of key importance in the debate over the impact of mining on Aboriginal people. After the Second World War exploration designed to increase Australia's self-sufficiency in aluminium led to the discovery of these bauxite reserves. They lay within the area covered by the Arnhem Land Aboriginal reserve. In 1963 the land was excised from the Aboriginal reserve and a final agreement on mineral development was reached in 1968. Production began in the early 1970s, and a new mining town, Gove (Nhulunbuy), was constructed. By 1990 that town, with a population of over 4,000, had become the fourth largest settlement in the Northern Territory. The mine, with an annual bauxite production of over six million tons, was then one of the largest in Australia.

The Nabalco development has, as Howitt (1992b) points out, been vital to the development of the Aboriginal land rights movement in Australia. About 25 km from Nhulunbuy lies the large Aboriginal community of Yirrkala, which together with its outstations, has a population of around 1,000. The protests of the Yirrkala people over the original agreement, which not only failed to allocate any compensation to them in the form of royalties but resulted in the loss of areas of land of great traditional significance, included a petition (the Bark Petition) to the federal government and a court case to have their traditional ownership rights affirmed. This was opposed by Nabalco on the grounds that the land was owned by the crown and therefore available for development. Although the Yirrkala people's claims led nowhere, they brought the land rights issue firmly into the realm of public debate, and arguably they were ultimately responsible for the passing of the Northern Territory land rights legislation in 1976. As Howitt (1992b) points out, it is ironic that Nabalco, because their agreement preceded this legislation, have been able to get away with paying relatively small royalty monies and the Yirrkala people, the first Aboriginal group to confront the powerful mining industry, have benefited far less than their fellows who did not have to deal with these issues until the 1980s. Not surprisingly, the Yirrkala people are now considering lodging a new claim on the basis of the Mabo judgment, in the hopes of receiving due compensation.

Economic impacts of mining

As suggested earlier, mining should theoretically be of great economic benefit to aboriginal people in remote areas because it alone has the potential to generate sufficient wealth to break the overpowering dependency on government funding. In practice its economic contribution is generally much less significant. It can in fact be negative. Following O'Faircheallaigh's framework (1991: 229–43) the economic impact of mining on aboriginal peoples is here examined under two main headings: the loss of other resources, including land; and the creation of additional economic opportunities through employment, the generation of cash from royalties and equity participation, and the provision of better goods and services and other infrastructure.

The loss of other resources

Mining development can clearly destroy other resources of economic value, particularly the land itself and associated soils, vegetation, wildlife and domestic stock. However, while the mines themselves and their associated infrastructure – seismic lines, roads, buildings, pipelines – are now an obvious feature of remote Canadian and Australian landscapes, the overall effect on the renewable resource base does not appear to be large. This is probably because many contemporary mining operations in these areas have developed only recently, and have therefore come into being at a time when Australian and Canadian based companies are required to take much more careful and stringent methods to prevent wanton and unnecessary destruction. It also reflects the general poverty of the rest of the resource base in many of these physically marginal regions.

Although the destruction caused by mining in either Canada's or Australia's north does not seem to be as widespread as has been reported for areas such as the Amazon basin, it undoubtedly has had a significant local environmental impact. This has caused much of the concern from aboriginal peoples. Specific fears have included that of direct contamination of water supplies, and the disruption of the subsistence resource base.

With certain types of mining, notably uranium in Australia's Alligator Rivers area, aboriginal people have been worried about contamination of water supplies and of the fish which form an important part of their subsistence base (AIAS, 1984). Strict environmental monitoring has been an integral part of the uranium projects at Ranger and Nabarlek where it has been conducted by the Office of the Supervising Scientist, a government body set up as part of the mining agreement (Fox *et al.*, 1977). No significant problems have yet been identified. However people are still not wholly convinced that the disposal systems used by the mining companies are safe because they are aware that, in these tropical wetlands with extreme monsoonal rainfall regimes, the possibilities for unforeseen events causing spills are still present. This happened at Nabarlek in 1981 following torrential rains associated with Cyclone Max (Tatz, 1982). Here an additional concern was the lack of publicity following the spill. As Tatz points out, because of the effects of such events on the whole subsistence food chain, it is essential that people are warned of any contamination problems which might have arisen.

Tailings threats to wildlife and water supplies are also discussed by Macpherson (1978a) in her analysis of the Cyprus Anvil lead/zinc mine in Yukon Territory and, in the course of Berger's (1977) inquiry into the construction of the Mackenzie Valley pipeline, many people expressed concern over the overall effects of mining activities on the aboriginal subsistence base. For example, in the Fort Norman area, adjacent to the oil and gas development at Norman Wells, people commented that caribou and moose were now harder to find, and that there were severe problems in setting traplines because of the proliferation of roads, seismic lines and drilling sites. Macpherson (1978b) also comments on the disruption to traplines caused by the Pine Point lead/zinc development, on the southern side of Great Slave Lake. People were also worried about how transport systems like roads or pipelines, or the inrush of large numbers of people with new and noisy technology might affect migrating wildlife. This is particularly important in Canada, where caribou herds are such a vital part of local subsistence economy. The conflict between oil development at Prudhoe Bay in Alaska and the Porcupine caribou herd has affected both Alaskan and Canadian aboriginal groups because the migratory range of the animals straddles the international frontier. A prime concern has been the proximity of the caribou calving grounds to Prudhoe Bay. It is unclear how they might be affected. Some researchers have noted that caribou in the Central Arctic herd near the trans-Alaskan pipeline have proved to be remarkably adaptable to changing conditions (Sheldon, 1988). However, as Russell (1988) points out, that herd has more widely dispersed calving grounds than the Porcupine caribou, and therefore can probably adapt more easily to disruption of their territory. Members of the Old Crow community in Yukon were clearly in no doubt about the problems which would arise for the Porcupine herd if Prudhoe Bay proceeded as planned (Yukon Government, 1988). They and others have argued that the risks are too great, particularly when the revenue from Prudhoe Bay might not be as great as anticipated.

The Nanisivik lead/zinc mine in Baffin Island provides another interesting example. Here, according to Dahl (1982), the mine had little direct environmental impact. However an important subsistence resource, whales and narwhals, was severely affected at a key period in the Inuit hunting calendar because the mining company used ice-breakers to open up their sea passage to Nanisivik at the earliest possible date after the winter freeze. This gave them a longer time for transporting ore to markets. Inuit in the neighbouring Arctic Bay community complained bitterly about this. Altogether, as Freeman (1985a) points out, the profound ecological effects, both human and environmental, of mining in areas as fragile as Canada's high Arctic must be taken into account.

The creation of additional economic opportunities

Employment

Common arguments favouring mining development in remote Canada and Australia are that these projects will generate significant employment for the existing resident population, including aboriginal peoples. Most evidence suggests otherwise. In 1982 in northern Australia, where Aborigines form 10 per cent of the population, they held only 2 per cent of jobs in the mining industry (Cousins, 1985: 80), and a 1987 study showed that in Western Australia the average aboriginal employment was less than 1 per cent (Rowe, 1989, cited in Howitt, 1992a). Examples of significantly higher levels of employment include the Comalco bauxite mine at Weipa in Cape York, where, as Howitt (1992a) records, the company has consistently employed Aborigines at rates of between 7 and 10 per cent of the workforce during the two decades since the early 1970s. This reflects recent Comalco policy which has encouraged local Aboriginal employment. This has not always been the case. In the early 1960s Aborigines were deliberately excluded from consideration for jobs at Weipa because it was felt that their labour was more vital to the pastoral industry (Stevens, 1981). Howitt (1989) records similar circumstances affecting aboriginal employment in the massive iron ore developments in Roebourne Shire in the Pilbara region of Western Australia.

Canadian evidence also shows low levels of aboriginal employment in mining, both for the Dene living in the vicinity of Norman Wells in the early 1970s (Watkins, 1977: 88–9), and at Nanisivik, where although the mining company had a policy of preferential employment for aboriginal people, only 25 per cent of the workforce were from that group.

Mining companies have generally failed to keep their promises to employ aboriginal people. They justify this by stressing that the contemporary nature of the industry with its overwhelming demand for a highly skilled professional workforce, precludes unskilled workers. Certainly this was ADM's explanation for the small size of their Aboriginal labour force, 10 out of 300 employees in 1987. It is also the conventional reason given for lack of aboriginal workers in any elements of mining which involve a degree of reponsibility. All of ADM's 1987 workers were in the 'site beautification' squad, responsible for planting trees to combat the ravages of erosion on the open-cast mine site. Tatz (1982: 173) also comments on the relegation of Aboriginal workers at Ranger and Nabarlek to rehabilitation work in the 'guise of employment'. He also points out that this is not a new development but was the result of an attitude common to mining companies in the 1960s and earlier, where Aborigines were seen to be inherently incapable of undertaking any but the most menial of tasks. As he suggests, this may well explain their apparent disinterest in working for mining groups, because the jobs which they are given confer no clear stake in the project and give them no lasting skills.

Direct aboriginal employment in mining could probably be increased through greater expenditure on training. However many mining companies are reluctant to take on these extra costs because they would inevitably detract from the overall commercial profitability of their enterprise. Some have taken their responsibilities seriously, although not necessarily specifically to foster local aboriginal employment. For example, between 1980 and 1986, Dome Canmar in conjunction with GNWT, Thebacha College and CEIC provided a training program at Tuktoyaktuk, the base community for their oil and gas operations in the Beaufort Sea. Esso Resources and Gulf Canada also later became involved and during the six years of its operations 'Tuk Tech', as the program was termed, trained northerners, both aboriginal and non-aboriginal, in a wide variety of skills including office work, industrial work and work concerning marine operations (Abele, 1989). Dome Canmar saw Tuk Tech as the means for achieving their goal of 20 per cent local employment in the workforce but, although they did provide special literacy support for their aboriginal students they still found that for many of them the chances of achieving the necessary levels of skill were poor (Dome, 1982). Not surprisingly, aboriginal employees have never accounted for a high proportion of the more highly skilled local labour force.

Both Tuk Tech and its successor Arctic College were run under non-aboriginal control – mining company and government respectively. Training has also been provided under partial control from aboriginal groups holding joint venture agreements with the mining companies. This should allow for more appropriate approaches to aboriginal training needs. Shehtah Drilling, owned 50 per cent by Esso Resources Canada and 50 per cent by Deh Cho Drilling Limited (Denedeh and Metis Development Corporations), was established in 1983 to train an aboriginal staff to work as oil-drilling crew and as company managers. It was also supposed to be a profit-making venture (Davies, 1984). As Abele (1989: 109–14) points out, these can be conflicting goals; spending resources on training undermines profits. Because of the highly competitive nature of the drilling industry, these conflicts escalated. At the beginning, aboriginal workers strove to acquire new skills because, above all, they valued belonging to an aboriginal-owned company. However, later, as the company paid greater attention to making profits, the desire for training dropped off.

Other problems highlighted by mining companies include high mobility and high absenteeism among the aboriginal workforce. This difficulty largely reflects the wide range of activities, both traditional and non-traditional, in which people are involved. Helpful remedies have included the rotation of jobs, discussed by Hobart (1981) for Coppermine in the Canadian Arctic, because people can move in and out of the mining workforce. However this clearly requires a flexible attitude on the part of the employer, and is bound to undermine the success of training programs.

Although comparatively few aboriginal people have actually been employed in the modern mining industry they have benefited from other employment spin-offs. Local business opportunities, such as working on contracts from the mining companies or providing goods and services, have also occurred. The example of Yirrkala Business Enterprises (YBE), an Aboriginal company concerned with a number of subsidiary activities associated with the Nabalco bauxite mine on Arnhem Land's Gove peninsula, is particularly interesting. The Methodist Church, which has always supported community development and self-sufficiency, was a strong force behind the establishment of the company in 1968, and until the late 1980s YBE's management was non-Aboriginal. YBE's attempts to gain its share of Nabalco contracts were only partially successful and in 1987, following the failure of real estate ventures in Darwin, the company faced debts of close to $700,000. Subsequently, as Howitt (1992b) summarises, the strength of Aboriginal control over YBE has been deliberately increased, with local Aboriginal management and all clans of the Yolngu people having equal interest in the company. Its activities now include Nabalco contracts for earthworks and rehabilitation, contracts with Nhulunbuy Corporation for beautification, garbage collection and other town services and a contract with the NT Department of Transport and Works for maintenance of the Gove–Bulman road. By early 1990 YBE's workforce, largely casual, had reached 170, compared to the total Nabalco workforce of around 820. Reasons for YBE's recent success include its emphasis on training, contract employment and on labour pooling, all of which are policies which accord with Aboriginal needs in terms of skilling and of flexibility of working conditions. This, as leaders of the Yirrkala community have stressed, has provided a much better working environment than that experienced by Aboriginal employees directly employed by Nabalco. It has certainly helped significantly in spreading the economic benefits from the Nabalco mine more widely within the local community.

Royalty payments and their equivalent

The receipt of payments through royalty agreements and their equivalents has contributed much more significantly to aboriginal economic development than has employment in the mining industry. In 1981, one group in the Alligator Rivers uranium mining area, for example, received royalty payments which averaged about $6,500 per head, when their average per capita income was only around $2,300 (O'Faircheallaigh, 1991: 238). This is obviously highly significant for them. However, these incomes are still not large enough to warrant the commonly held assumption that Aborigines in that region have become 'uranium sheikhs'.

Amounts paid through royalty payments are very uneven, both because of variations in the agreements through which they flow and because of fluctuations in the profits of the companies concerned. Their subsequent uses, whether for individual or common benefit or for investment or immediate spending, also vary markedly.

For many aboriginal people lack of land rights legislation, coupled with lack of bargaining experience and political power on national and international scales, has excluded them from royalty negotiations. In fact it is only during the last two decades that, through land rights recognition, some aboriginal groups have been able to gain substantial cash benefits via these channels. Australia, with land rights legislation which differs on state, territory and federal bases, provides particularly good examples of such variations between royalty arrangements. These essentially split Aboriginal groups affected by mining into those who 'have' and those who 'have not'.

Since 1976 Australia's Northern Territory Land Rights Act has given Aborigines significant control over exploration, mining and the revenues from sub-surface resources on their land. Thus by the end of 1985 the Gagadju Association, the local group adjacent to the Ranger uranium mine, had received over $12 million in royalties. They were free to use these as they wished, for investment, capital expenditure or individual purposes (O'Faircheallaigh, 1986b). In the Tanami desert in central Australia the Janganpa Association, representing the traditional owners in the area of the North Flinders gold mining development, received almost $5 million in the five years from 1984 (Howitt, 1991). In contrast royalties received from the Nabalco mine at Gove have been much smaller, and have included no negotiated component. Here the agreements, negotiated long before the introduction of land rights legislation, excluded the Aborigines. In 1992 Yolngu leaders, through the Northern Land Council, continued to give notice that they were seeking to renegotiate the 1968 Gove Agreement, to gain a fairer distribution of benefits for the local Aboriginal group (Howitt, 1992b).

Outside the Northern Territory Australian Aborigines have been much less successful in gaining economic benefits through royalty type agreements. This is largely due to inadequate recognition of land rights. Traditional owners in the East Kimberley region of Western Australia, for example, were initially granted $300,000 per annum through the Good Neighbour Policy (GNP) agreement signed with the Argyle Diamond Mine consortium. This money was distributed to only three Aboriginal communities, and many others with interests in the barramundi dreaming site destroyed by the mine received no cash compensation whatsoever. Moreover the administration of

Plate 5.3 Open-cast mining at Argyle: excavation of the kimberlite pipe, location of the barramundi dreaming site

the funds left control in the hands of government and ADM and restrictive clauses in the agreement prevented the Aboriginal communities from using the cash as they wished. They were allowed to spend it on buildings, infrastructure, vehicles and station improvements. They were not, however, allowed to invest it or use it to buy more land. Comments on the paltry nature of the payments and the unfairness in their distribution led to a revision of the arrangements and in 1984 GNP was replaced by the Argyle Social Impact Group (ASIG). For the next five years this organisation administered a fund of $1 million per annum, contributed to on a 50 per cent basis by ADM and the Western Australian government. ASIG payments could be made not only to the existing GNP communities but also to other East Kimberley groups. Although the fund was larger, many of the restrictions remained. People were still prevented from buying additional land, the prime need of most people in the region. In 1989 ASIG was wound up, and compensation from ADM reverted to the earlier GNP agreement. Once again, only three communities benefited. Altogether it seems that, while communities near Argyle now undoubtedly have better housing and services than before mining, the long-term economic advantages to the East Kimberley people have been very small. Considering the value of the mine (reputedly with gross production worth around $500,000 per day in the first year of its operation, and subsequently averaging $394 million annually) the amounts accruing to the East Kimberley Aborigines are grossly insufficient (Christensen, 1985; Dixon and Dillon, 1990).

Comalco's arrangements over the development of their Weipa bauxite deposit in Queensland's Cape York region provide an alternative approach in a region where there is no legislation promoting royalty agreements. There Comalco, state and federal governments initially allocated $2 million over ten years for improvement of community infrastructure and facilities in the neighbouring Aboriginal community. Between 1973 and 1990 the total amounts allocated had exceeded $3.7 million (Howitt, 1992a). These monies have been administered through a trust body, the Weipa Aborigines Society, which has increasingly come under Aboriginal control. As a result the expenditure has moved from an emphasis on infrastructure such as sewerage, roads and drainage and improvements to a community hall to an emphasis on training and development, not specifically related to the mining industry. This has included enhancing people's skills in dealing with their major education and health problems, including alcohol rehabilitation, and their skills in general areas of community management. As Howitt (1992a) points out, this is in direct contrast with ADM's Good Neighbour Scheme where no such types of expenditure have been condoned, and the emphasis is placed firmly on conspicuous forms of development such as new buildings.

Canadian comprehensive land claim agreements also contain clauses granting sub-surface rights to certain negotiated areas. Consequently any mining occurring in these regions would involve royalty payments. As yet this has not led to large payments of cash to aboriginal communities, principally because only two such agreements (James Bay and Inuvialuit) are currently in place. Under the latter agreement Inuvialuit people gained outright mineral rights to 15 per cent of the 90,000 sq. km to which they gained title in fee simple absolute. But the area granted to them excluded those regions with proven oil and gas reserves at the time the agreement was signed in 1984. Their compensation payment of over $150 million was intended partly to offset this. The aborted Dene/Metis claim for the Mackenzie Valley region included similar provision ($70 million) to compensate for the exclusion of the Norman Wells region with its proven oil and gas reserves (Young, 1991c).

Royalty monies need not only benefit the traditional owners of the land around the mine. They may also benefit others. Under Australia's NT Land Rights Act (1976) mining monies are paid into the Aboriginal Benefit Trust Account and then divided into three components: payments to traditional owners (30 per cent); payments for the support of the NT Aboriginal Land Councils (40 per cent); and payments into a fund for disbursing grants and loans to Aboriginal communities throughout the territory (30 per cent). By 1991 this fund had received $230 million in total, primarily through payments from uranium mining; over $47 million had been disbursed as grants to assist a wide range of projects in both large and small communities in all NT regions. In the early to mid-1980s significant amounts of this money went for purchasing vehicles for people re-establishing themselves in outstations on their country. More recently ABTA has also been used to buy four Northern Territory pastoral stations, thus augmenting the federal government's funding allocation for land acquisition.

Equity and joint venture agreements

Aboriginal people in remote regions can also benefit through direct participation in the mining venture, hence sharing in the profits. This normally involves investment through a development corporation, perhaps as part of a land claims agreement. In 1987, for example, the Dene/Metis Fort Good Hope Development Corporation in the Mackenzie Valley signed a joint venture agreement with Chevron Oil for oil and gas exploration in the vicinity of their community. Along with compensation for detriment caused to land and resources through Chevron's activities, and representation on environmental impact committees, this agreement included a 20 per cent option on licences arising from the discovery of minerals . Fort Good Hope also set up a gravity and survey company employing 16 people, and seasonal jobs on the venture, numbering around 100 in 1987/88, were expected to increase rapidly. Such arrangements clearly give aboriginal people much more control over vital elements of the industry, both on how and where it develops, how its profits are distributed in relation to those goups most obviously affected, and in areas such as employment and training. But, as O'Faircheallaigh (1991: 241) points out, there are risks involved. In particular, the venture may, for any number of reasons, fail. That is a risk which has to be taken. However the risks can be minimised. In the Fort Good Hope case, Chevron Oil apparently not only invested significant resources in training schemes related to the venture but also assisted the community in ways which might have longer-term value – helping the Hunters and Trappers Association with the expenses of provisioning the winter hunting season; and assisting with drug and alcohol programs to help people to cope with some of the impacts of increased incomes.

Canadian comprehensive claims, all of which include substantial payments in compensation for loss of rights over land, provide obvious potential for aboriginal groups to buy into joint ventures and equity arrangements. Development Corporations established under these claims settlements, such as the Inuvialuit Development Corporation (IDC), would have the option to share in mining development on their own lands, possibly by becoming part of the mining consortiums involved. Oil exploration, development and production companies, and related activities such as contract drilling services, oil-spill response capabilities and surveying and mapping services have been among the investments made since IDC's establishment in 1984 (Robinson et al., 1989a). However, as commentators have pointed out (for example, Pretes and Robinson, 1989a; Robinson et al., 1989a; Robinson et al., 1989b), while this gives the Inuvialuit an important stake in the industry, all-out investment in large-scale mining and non-renewable resource developments is not necessarily a good strategy. Not only is it risky because of uncertainties inherent in these industries, but it does little to create employment and keep cash within the local community. For example, Polar Gas (1985) estimate that the whole Mackenzie Valley gas pipeline from Inuvik to Northern Alberta would in the end provide only 213 permanent jobs, of which one-third would require such high levels of skill that they would almost inevitably be filled by people from outside the region. In around thirty years' time, when the pipeline ceased to operate, these jobs would be gone (Pretes and Robinson, 1989b). An additional problem is that the local aboriginal people have very high expectations of investment corporations such as IDC, and anticipate that they will not only generate cash, but that they will then distribute it within the aboriginal community, either to support local business enterprises, or through dividends, or through the creation of more jobs. Pretes and Robinson (1989b) suggest that these Development Corporations should not restrict their interests to non-renewable resource developments but should also invest in a variety of other types of business. These might include projects concerned with service provision or with renewable resource processing, many of which might have much more to offer local aboriginal people in the long term.

Infrastructural improvements

Mining can also lead to improvement in communication systems and essential infrastructure throughout the whole region. The development of Norman Wells was followed by the establishment of winter 'ice roads' which have benefited aboriginal communities in the upper Mackenzie Valley. The establishment of new ports, as at Groote Eylandt for the manganese mine and new airstrips, for example at Tuktoyaktuk, have allowed aboriginal people in nearby communities to obtain cheaper freight rates on goods and equipment (O'Faircheallaigh, 1991). New infrastructure can also help the development of traditional activities. Aborigines wishing to establish outstations in their traditional country in the remote north west corner of South Australia have felt more confident about travelling through that area after BHP exploration teams constructed rough roads, and have said that they value the assistance which the miners have given them when their vehicles break down. These examples suggest that the infrastructure development needed for mining can have spin-offs for aboriginal peoples. However it must be stressed that this can only happen if aboriginal people have access to that infrastructure. As we have already seen, many large mining companies form completely separate entities, and outsiders are not free to use whatever services or facilities they may have established.

The use of cash from mining: positive and negative economic effects on aboriginal peoples

The above discussion has shown that mining can indeed bring significant amounts of cash into the local aboriginal economy. These payments should, considering the poverty of that economy in monetary terms, make a significant impact on community and individual incomes. The evidence has also shown that in both Canada and Australia the actual amounts of cash received through mining vary markedly in relation to population. Many aboriginal groups have received virtually nothing to compensate them for the loss of their land and its natural resources, or for the disruption to their lifestyles caused by the intrusion of non-aboriginal communities of very different cultural and economic aspirations. For them the economic impact of mining is either absent or negative. For those who do receive cash, the way in which that cash is used becomes a vital issue in terms of development. This is particularly the case with royalty payments, or profits from joint venture/ equity enterprises, where the sums available may be quite large.

O'Faircheallaigh (1991: 241) comments that until now there has been remarkably little research on the factors which affect the use of royalties and other large sums of money from mining. As he points out, this is a complex area of understanding, combining knowledge of legal and institutional structures with social and political factors affecting the circumstances of different aboriginal communities or individuals. His own work with the organisations dealing with the payments to two of the Alligator Rivers communities in Australia's uranium provinces, the Gagadju and Kunwinjku associations, illustrates some of the main variations which occur. By 1985 the Gagadju Association had used over 50 per cent of their income to purchase considerable assets which would provide them with additional income both for the present and in the future. This included buying and further augmenting tourist facilities in Kakadu National Park, situated on Aboriginal held land adjacent to the mine site; they also improved their own educational and health services and helped those members who wanted to establish outstations back on their traditional lands (O'Faircheallaigh, 1986a; 1988). Only 13 per cent of the income was distributed to Gagadju Association members. In contrast the Kunwinjku Association spent significant sums on buying vehicles which, through heavy use, only lasted for a short time. They also suffered some substantial losses through unfortunate investments and dishonesty among employees. Altogether, when the Nabarlek uranium mine ceased to function in 1988 most of the traditional owners of the land had very little to show for the monies disbursed (O'Faircheallaigh, 1988).

Choosing the appropriate mix of expenditure between investment and distribution of revenue to individual claimants is not easy. Robinson *et al.* (1989a) suggest that a high proportion of these sums should be invested. However people are also very keen to receive cash in hand. Indeed if they do not, they may well accuse their associations of financial mismanagement. Where individual payments are made there is of course no control on how that money is eventually used, whether for buying alcohol or gambling; or for purchasing equipment such as rifles, vehicles or outboard motors; or simply for day-to-day food and clothing purchases. Regardless of what is decided, the actual amounts involved are unlikely to make any individuals rich by conventional non-aboriginal standards.

Social impacts of mining

Social impacts of mining include: the social and cultural effects of loss of land and its renewable resources; the effects of introduced communities with populations which are culturally very different from resident aboriginal peoples, and which are materially much better off in every way; and the social effects of increased cash incomes on the aboriginal community.

Loss of land and resources

The loss of land and resources through mineral development can have very severe social consequences for aboriginal people. Both, as has already been discussed, are highly valued for social as well as economic reasons and destruction of the land, particularly if it involves places of great spiritual

176

significance, is tantamount to destruction of ancestral beings. Confrontations of this type have been particularly marked in remote parts of Australia. The whole sorry saga of ADM and the barramundi dreaming site (see p. 158) has, as Dixon and Dillon's (1990) collection of papers illustrates, not only caused conflict between Aborigines and miners but also led to marked rifts within the Aboriginal community over who had the right to talk about the site itself. Today, with the kimberlite pipe which forms the site all but gone, people are resigned to its loss. But they are still gravely concerned both for themselves and for future generations, seeing the loss as evidence of their own failure to preserve the land for their children and grandchildren (Ross and Bray, 1989). In Western Australia, with its lack of land rights legislation, agreements such as that accepted at Argyle were possible without adequate environmental and social impact assessments because Aboriginal opposition was restricted. More recent opposition to the Coronation Hill mining development in stage 3 of Kakadu National Park in the Northern Territory has been sufficient for the federal government to refuse permission for CRA to proceed. At present this rainbow serpent site is safe. Whether it remains so depends not only on Aboriginal feelings, but also on external political factors such as trade-offs between the federal government and the Northern Territory government, almost invariably in favour of mining whatever the consequences; and economic factors such as the price of gold, platinum and palladium, the combination of minerals in the deposit.

In-migration and the establishment of new or expanded non-aboriginal settlements

The inevitable influx of population associated with mining projects has obvious social consequences for the resident aboriginal populations. The cultural and social characteristics of the newcomers, their lifestyles and expectations have generally been very different from those of aboriginal people and resultant misunderstandings have led to conflict initiated from both sides. Social problems for the aboriginal groups have also arisen because some of them, often the younger and more highly educated individuals, have wanted to adopt many of the practices and advantages which they perceive the newcomers to have, and this has disrupted traditional forms of social control. They have also arisen because many of the incomers live in isolation from their own family groups and seek relationships with individuals in the aboriginal community in ways which are rarely permanent and usually exploitative.

Conflict over the use of local natural resources has arisen in both Canada and Australia. Berger's (1977) inquiry cited a number of comments over game hunting and sports fishing carried out in the Mackenzie Valley by mining employees in areas where such resources had always been used and controlled by aboriginal peoples, for whom they formed a vital subsistence base. In Australia, because of the absence of cash enterprises based on such resources, conflicts of this type have perhaps received less attention. They have still occurred, even when the mining companies have tried to prevent them. In the development phase of ADM conflicts arose at favourite fishing holes on the Ord and Dunham Rivers, and on one occasion a group of miners dynamited a waterhole, killing not only all the fish stock but also crocodiles and other aquatic species. Although ADM immediately dismissed these employees, the Aborigines remained very concerned that such events could occur again. In coastal areas in the vicinity of Nhulunbuy there has been conflict over recreational four-wheel driving over sand-dune areas by mining employees (Howitt, 1992b). Such problems reflect not only different perceptions on the use of the land and natural resources. They also reflect the severe restrictions that many mining company workers face during their sojourns in these isolated communities. Not only are they often cut off from family and friends, but if the mine is on Aboriginal land, they are surrounded by country to which they may have very limited access. The Ranger Uranium town at Jabiru, for example, is entirely surrounded by Aboriginal land, designated as national park and administered by the Australian National Parks and Wildlife Service. They can use Kakadu in the same way as any other tourist, which means obeying restrictions on access and activities such as fishing; but if they want to range more freely they will have to go elsewhere, inevitably some distance away. Nhulunbuy similarly is surrounded by Aboriginal land. Clearly the types of conflict arising in such situations require some compromises on both sides.

The new mining towns have also attracted aboriginal people, not necessarily seeking employment but often because they want to use the services, or merely because of the 'bright lights'. Since these towns are not set up to cater for such groups, especially in terms of shelter, aboriginal in-migration could lead to the growth of informal settlements in the surrounding area. Mining companies generally regard the growth of 'humpy camps' on their outskirts with great trepidation, and take strong measures to prevent it happening. Strict supervision of entry to the town, as at Argyle, is one measure; constant surveillance of the movement of aboriginal people around the mining area is also common. This can be extremely intimidating, as I have discovered when transporting Aboriginal groups in the Argyle area. The sight of an ADM vehicle was usually enough for people to ask me to take a wide detour. Since the company owns the settlements and takes responsibility for providing services only for their own employees, such restrictions are perhaps understandable. However it does artificially isolate aboriginal and mineworking groups from each other, and obviously hinders the growth of better social relationships between them. This need not always be a disadvantage. But it does little to promote any cross-cultural understanding which could be of long-term benefit.

The isolation of single mineworkers, mostly male, has inevitably led to liaisons with aboriginal women, and to an increase in the number of part-aboriginal children in local families. This can have both social and economic consequences. Howitt (1989), for example, points out that in the Roebourne area mineworkers rarely provided sufficient financial support for their Aboriginal children, particularly after they had left the district, and that the financial burden on the poorly endowed Aboriginal community was quite considerable. Tatz (1982) notes concern in the east Arnhem Land communities about what will happen to the 'Kids That Are Not True', those that belong to two cultures but whose futures may inevitably restrict them to only one.

Aboriginal people are well aware of these problems and in many cases have openly voiced a concern and indeed fear of the consequences (see, for example, Berger, 1977; Fox *et al.*, 1977). Many see the intrusion of outsiders as basically disruptive, not only because of conflicts over the use of land around the mine but also for less tangible reasons such as the disruption of social and cultural structures. Tatz (1982) comments that in Arnhem Land people were particularly concerned in case the uranium developments disrupted the ceremonial linkages between communities in southern Arnhem Land and those directly affected by the mines. Recognition of these problems, coupled with the problems faced by mine-workers and their families forced to live in such isolated situations has led mining companies to adopt alternative forms of settlement involving fly-in/fly-out systems. In Australia the Nabarlek uranium mine functioned entirely on that basis during the ten years of its existence, and the fact that miners were only present while on shiftwork and undertook their recreation in Darwin probably made local conflict over use of surrounding areas of Arnhem Land less likely (Tatz, 1982: 179–80). Both the ADM and the North Flinders Tanami gold mine operate on this basis. In Argyle the main workforce of around 450 operates in two fortnightly shifts, transported direct from Perth to the site as single workers. Company managerial staff are housed in Kununurra and commute the 150 km to the mine each day by air. North Flinders Mines similarly transports its 300 or so workers in shifts from Alice Springs. Mines such as these in some ways have more in common with off-shore oil rigs than they do with conventional mining towns of earlier periods, such as Weipa or Nhulunbuy in Australia or Yellowknife or Norman Wells in Canada.

The effects of increased access to cash

The attractions of the mining towns for aboriginal people include not only better facilities such as well-equipped schools and hospitals but also well stocked supermarkets and all sorts of recreational activities. These attractions are linked to the fact that many aboriginal people living near mining operations have far more ready cash than they have ever had in the past.

Consumerism, in a whole variety of forms, inevitably becomes important. These have included increasing use of alcohol/drugs and consequent increase in violence, suicide and other behavioural problems; a desire for more consumer goods, particularly vehicles; nutritional changes caused by increasing dependence on store-bought processed foods; internal social conflict caused by inequalities in the distribution of the monies; and the creation of a dependency on a source of wealth which may well be ephemeral.

Alcohol and related problems

People living in Canadian and Australian remote area towns, such as Yellowknife, Inuvik, Darwin and Alice Springs, in general have a reputation for high alcohol consumption, enshrined in songs such as 'Bloody Good Drinkers in the Northern Territory' and borne out by figures which show that beer consumption in the Northern Territory in the late 1970s was 70 per cent higher than for Australia as a whole (AIAS, 1984: 208). This reputation is particularly associated with the mining industry, the 'boom' and 'bust' economy which brings in large numbers of temporary workers. It is not a recent phenomenon, as descriptions of the rip-roaring life in goldrush days in the Yukon or in old Halls Creek will testify. Aboriginal people in both Canada and Australia were for long debarred from taking part in these activities, and it is only during the last three decades that alcohol prohibitions have been officially abandoned. Alcohol abuse has subsequently been recognised both by non-aboriginal groups and by aboriginal communities and organisations as a severe problem for many people. It is a clear sign of social and cultural disruption arising from the impact of external development on small communities not equipped to deal with its effects. As Australia's recent inquiry into Aboriginal deaths in police custody has shown (RCIADIC, 1991), it has other tragic consequences: violence, suicide, accidents and alcohol-related disease. However, as O'Faircheallaigh (1991) rightly points out, it is difficult to decide whether or not these problems can be attributed mainly to the effects of large-scale resource development such as mining. Drinking habits have often been firmly entrenched before the mining operations began. It is perhaps the sheer scale of mining developments and the rapidity of their growth which has made many people feel that they are a major factor in excessive alcohol consumption.

Evidence for alcohol problems arising through contact with mining operations certainly does exist. For example, the Ranger Uranium Environment Inquiry (Fox et al., 1977) foreshadowed alcoholism as one of the main problems which Aboriginal people would encounter as a result of the Jabiru and Nabarlek developments. However it is hard to assess the extent to which this has occurred. In Oenpelli, the nearest large Aboriginal community to both mines, as Tatz (1982) points out, there was a history of grogging in the community before mining. Until 1981 alcohol was not legally allowed within the Arnhem Land reserve, including Oenpelli. In that year this legislation was revoked and the community then successfully applied to be declared 'dry' under the NT Liquor Licensing Act. That effectively meant that Oenpelli 'drinkers' either had to use the community beer club or had to have a permit to drink. Tatz (1984), comparing Oenpelli with other large Aboriginal communities operating under similar systems, commented that since Oenpelli issued a comparatively small number of drinking permits many people must have found the local club sufficient for their needs. Thus, although alcohol consumption probably increased following the development of Ranger Uranium, that increase might not have been substantially higher than had occurred in other Aboriginal communities in the 1980s.

Although excessive drinking does not normally affect whole communities, the small numbers of people involved can, through other antisocial behaviour, have a disproportionate effect on others. Kesteven (1984) comments on the fact that break-ins to premises thought to have alcohol, such as the club, were usually the work of a known Oenpelli drinking 'gang'. Other major costs to the community, apart from the obvious cost of the alcohol and the extent to which this takes away cash incomes which are desperately needed to feed other members of the family, include: violence – suicide, domestic violence and fighting between rival groups; and neglect, both of other family members and of responsibilities, including those associated with ceremonies.

Alcohol has also been identified as a major problem associated with large-scale resource development in northern Canada. It was mentioned by many aboriginal witnesses in the Berger inquiry (1977) and subsequently caused concern for Dene/Metis communities associated with the Norman Wells pipeline development in the early 1980s. The Dene Gondie study, a community development research project administered by the Dene Nation in conjunction with a team from the University of British Columbia School of Community and Regional Planning (Rees, 1986), has stressed the importance of this issue. Many Mackenzie Valley communities involved in the government-funded Dene Community Development Program, part of the impact monitoring system for Norman Wells, established their own alcohol rehabilitation projects because they saw these as a prime need (Gorman, 1986).

Increased desire for consumer goods

Mining monies have undoubtedly allowed people not only to buy more consumer goods, but also to buy items which are both larger and more costly than in the past. This has had both advantages and disadvantages. Material living standards have risen and many people are better dressed and equipped than they were before. O'Faircheallaigh (1991: 246) cites comments from Aboriginal women in Oenpelli about their pleasure in being able to afford washing machines and refrigerators.

The purchase of vehicles with mining money has become an important issue in itself. Advantages include having better access to important services, such as health (O'Faircheallaigh, 1991: 246), and being able to travel more easily to see family and friends. But it can have tragic consequences, most obviously when linked to alcohol consumption. In the late 1970s and early 1980s vehicle accidents involving alcohol consumption increased substantially around Oenpelli, with over 40 per cent of accidents between July 1979 and May 1981 falling into this category; two deaths and seven injuries occurred during this period (Tatz, 1984). Tatz suggests that this had occurred partly because the Oenpelli community had successfully opposed the granting of take-away liquor licences throughout the surrounding district, and as a result committed drinkers were forced to drive far afield to buy grog. This would obviously increase the risk of drink driving.

Purchased food

Having more money has also allowed people to buy more of their food, and also to purchase items which formerly might well have been too costly. On the whole most people would see this as a distinct benefit, and certainly the ability to include fresh foods in the diet, always an expensive business in the outback, must have had positive effects on people's health. Disadvantages include the replacement of bush tucker and country food with these items. Foods from the land are often highly nutritious in mineral terms and provide good sources of protein and fibre; 'fast foods' from the store are not so good. Schaeffer (1983) records an increase in obesity and in associated diseases such as diabetes in an Inuit community affected by oil development, and similar dietary changes have certainly been occurring in some Australian communities.

Internal social conflict

Inequalities in the distribution of the cash and in access to prized objects such as vehicles have undoubtedly caused social conflict both within and between aboriginal communities. Other conflicts have been related to different interests in land, and different perceptions of the relative worth of traditional values compared to those of modern life.

As Australia's Argyle Diamond Mine has demonstrated, taking the initial decision to go ahead with mining, and agreeing to the terms for development are actions fraught with difficulty. The agreement signed by traditional Aboriginal owners of the barramundi site at Argyle was not only completely inadequate in terms of the returns to disadvantaged local Aboriginal groups. It also, according to other Aborigines, did not take the views of the whole community into account and excluded the signatures of key people (Christensen, 1985). CRA's swift action in transporting six Aboriginal leaders to

Perth to sign an agreement never subsequently made public, caused dismay and anger within the Kimberley Aboriginal community and has had long-term effects in undermining the local political authority of those involved. It has been suggested that CRA applied unacceptably strong pressure on signatories, promising them the necessary financial support to establish their new outstation at Glenhill in exchange for permission to proceed with the mine. While other Aboriginal people recognise the pressures to which they were subjected, the bitterness remains.

Subsequent events at Argyle also illustrate how mining has caused social conflict. As discussed earlier, the initial agreement applies to only three East Kimberley Aboriginal communities, who between them will share around $300,000 per annum for the entire life of the mine. Between 1984 and 1989 this agreement was extended with West Australian government assistance to produce the $1 million per annum of the Argyle Social Impact Group, a fund available for distribution to all Aboriginal groups in the area. While this did help to distribute some of the benefits more widely, the basic discrepancies remain and there has undoubtedly been jealousy between the original recipient groups and the others. As those forced to live in camps on the periphery of Wyndham or Kununurra point out, they are doubly disadvantaged because they have already been forced off their land by the cattle industry. Now their traditional interests in that land are being ignored.

Further conflicts concern the disruption of traditional systems of authority when some people are determined to preserve all places of spiritual significance and prevent mining at all costs; and others feel that the benefits of the money, jobs and other advantages which might flow to them are too great to be ignored. To some extent this conflict corresponds to age groups, with the former group usually consisting of older people and the latter the younger people with more formal experience of education. However, that division is not completely clear cut and, as more recent comments made by Aborigines associated with the Northern Land Council show (Howitt, 1992b) more people are coming to recognise that mining, in the right place, can be acceptable. The key issue is aboriginal control over deciding where mining can take place and how its benefits will flow through to the community.

The ephemeral nature of mining

Finally, mining is often only a short-term development. Receipts of cash from mining, jobs, improved infrastructure and other material benefits are obviously of value. But, as some aboriginal people well recognise, they create a demand for higher standards of living which may be very difficult to maintain once the life of the mine has finished. Moreover, in the process, some components of life, particularly the cultural component, may be irreparably damaged and even lost for ever. The Dene Nation, very

183

concerned about the effects of wage labour and relatively high earnings on their young people working in the Mackenzie Valley oil developments, put forward a cultural survival program as a keystone of ongoing community development (Dene Nation, 1984). This centred particularly on providing proper training and support for young people so that they would know how to live from the renewable resources of the land. It also involved producing appropriate material for use in local broadcasting media to counteract what were seen to be the destructive effects of conventional television and radio. As with other social pressures such problems are not solely caused by mining activities; but mining is, in remote areas, often the focus because it is the reason for the growth of new radically different human settlements.

Argyle also provides a clear example of the problems arising from dependence on the proceeds of mining. The distribution of cash allocated through the Good Neighbour Program has been administered by ADM employees . The GNP communities have received new housing, new school and new clinic buildings, fencing and yards, water tanks and vehicles. All of these have been extremely useful. But these were not the only things that people needed. Measures designed to promote employment and training might well have been preferred. Furthermore, all this infrastructure paid for through GNP has to be maintained, and there may well be problems with finding the funds for that in future. A further problem, commented on by a number of people (Dixon and Dillon, 1990: 119; Coombs et al., 1989), is that conventional government funding agencies have deliberately reduced their own assistance to these groups because they know that they receive GNP money. Their excuse is that funds are scarce anyway and that others have greater need. The overall effect has been to force these three communities to become increasingly dependent on ADM. Amongst other things this makes them vulnerable to pressures from the company, particularly if there are any further discoveries of rich diamond deposits in the area. And once the mine closes, what will happen to them?

Aboriginal people and mining: the future

As the above discussion shows, differences in the approach to remote area development on the part of the mining industry and the aboriginal people who live there are substantial. They include concepts such as short-term as against long-term benefit; large-scale as against small-scale operations; capital intensity and high profit generation as against labour intensity and a stress on maintaining human values; and above all a view of the land as a repository of resources to be tapped, as against the land as the spiritual and cultural basis of society, within which resources must be husbanded for future use. It is scarcely surprising that, with such differences in approach, conflict between aboriginal views and those of mining company employees has been common. Yet mining has not been totally detrimental to aboriginal interests. Through carefully negotiated agreements, which confer meaningful control on those with traditional rights to the land under discussion and which ensure that benefits from the mining venture flow through to those most affected, mining can make extremely valuable contributions to the process of self-determination, and can even be a key factor in supporting other activities to which aboriginal people are more committed. This can even include the development of renewable resource use. It is therefore clearly important that such negotiations occur. A major issue here has been the acceptance, on the part of both governments and mining companies, that social impact assessment will be part of the initial negotiation procedures.

Both Canada and Australia now provide some important examples of how social impact studies have helped to raise awareness on the part of both aboriginal people and mining developers of their respective needs and attitudes. These studies include, in Canada, the Berger inquiry into the Mackenzie Valley Oil and Gas Pipeline and more recently the Dene Community Development Program and Dene Gondie studies into the impact of further development, including a pipeline from Norman Wells. In Australia they include the Uranium Social Impact study and the East Kimberley Impact Assessment Project; more recently they have also included components of the Resource Assessment Commission's inquiry into the proposed venture at Coronation Hill. These studies range widely both in the scale of their operations and in the way in which they are administered and conducted. It is worth a brief summary of some of these studies to consider their contribution to solving potential areas of conflict, both at present and in the future.

In many ways the Berger inquiry (Berger, 1977) has set the scene for all subsequent investigations. Although originally set up as an examination of the environmental and economic feasibility of constructing a pipeline to bring natural gas from Alaska's Prudhoe Bay field through the Mackenzie Valley to southern Canadian and United States markets it was broadened to include a detailed study of the project's social impact. It collected evidence not only from a vast range of practitioners in the environmental arena and from key people in the mining industry but also from residents of every part of the vast Mackenzie River Valley. Most of the latter were aboriginal people, and as many indicated, this was the first time that their views on what was happening in the development of their country had been actively and publicly canvassed. As their comments indicated, they had many immediate concerns over their social and economic condition and requested not only assistance to deal with these but above all time and space to come to grips with the demands being made of them by the rapidly changing world around them. The prime result was the delay in the construction of the Mackenzie Valley Pipeline and agreement for further negotiation and consultation over land issues before that should occur. While many people certainly disagreed with the study, and it has

185

subsequently been criticised by academics and others (Asch, 1982), its impact on the understanding of many Canadians and indeed of people in other countries was highly significant. The issue of resource development and aboriginal peoples in remote areas was placed firmly on the agenda.

Subsequent government-funded studies, like the Berger inquiry, include Australia's Uranium Social Impact study in the Alligator Rivers area of west Arnhem Land and Canada's Dene Community Development Study of communities around Norman Wells. Although both of these owe their philosophies partly to the Berger inquiry, they have been conducted in totally different ways.

The Alligator Rivers Uranium Social Impact Study stemmed directly from concerns voiced in the Fox report and subsequently taken up by the Northern Land Council. These basically expressed the fear that the large influx of people, principally non-Aboriginal, into the area would threaten Aboriginal welfare and interests, cause tensions and conflicts and lead to Aboriginal resentment of the higher living standards of the incomers. Social monitoring of these situations was funded through the federal Department of Aboriginal Affairs which allocated around $750,000 for a five year project beginning in 1978. Because the other monitoring agency for the mining development, the Office of the Supervising Scientist, was specifically concerned with environmental issues the social monitoring project was taken on by a different organisation, the Australian Institute of Aboriginal Studies. During the five year period the researchers worked very closely with the communities most heavily involved, primarily Oenpelli and its outstations, situated between the two mines. Their study was a detailed anthropological investigation, covering an extremely wide range of issues including legal and administrative structures in the community, the economic consequences of mining, health issues related to uranium contact and to alcohol and the interaction between the two 'separate' societies, Aborigines and non-Aboriginal miners.

In comparison to Berger's inquiry, this study was very much smaller in scale and dealt with conditions and results which were specific to one particular community. Fieldwork was undertaken at much greater depth and the involvement of the researchers in the community was obviously much greater. Many of the issues raised by the study – the positive and negative economic effects, the problems hindering Aboriginal participation in employment and training and the alcohol problem – pointed to a 'Society in Crisis' (AIAS, 1984: 299), and these, plus the recommendations for delay in further development until such problems had been considered in greater detail, were similar to those raised by Berger and were of much broader relevance.

The Dene Community Development Program was a major component of a $21 million federally funded program designed to establish means to lessen the negative impact on local aboriginal groups of the Norman Wells Pipeline and Oilfield Expansion Project, approved in 1981. During the four years of its operation under the Dene Nation, the main representative aboriginal body in the area, it attempted to provide the 26 communities within the region with human and financial resources which would allow them to identify their own needs, concerns and impacts (Gorman, 1986: 11). These records would then form the basis of specific projects of long-lasting benefit. Issues raised by the communities included alcohol/drug abuse, economic development, health and education and the need for leadership and professional training and major efforts in training young people in the necessary skills for living off the land were made by 20 out of the 26 groups. Unfortunately, although workshops to discuss these issues and make plans for the next stage were held, the project had to disbanded in 1985 because funding ceased. This followed the allocation of the $9 million costs for the training expenses of CEIC to the program, and the consequent total allocation of the budget. The Dene were gravely disappointed, both because the program had been going well but also because their efforts had from the beginning been hampered by delays in the release of funds and by problems encountered in meeting the bureaucratic regulations required for release of money (Rees, 1984 and 1989).

All of these studies were government funded, with funds to a large extent controlled by outside agencies. But they all exhibit a strong degree of community involvement and present findings which are firmly based on community perceptions of their needs and wants. Other social monitoring projects have gone somewhat further, not only involving co-operative and collaborative research with the community but also conducted with resources from outside the government sector. In these cases control lies more clearly with the aboriginal people. Such studies have resulted from careful negotiations between aboriginal groups and interested experts, usually academics with whom long-standing contact and trust had been developed. The Dene Gondie study and EKIAP fall into this category.

Dene Gondie, a co-operative research project run by the Dene Nation and the UBC School of Community and Regional Planning, was designed to monitor socio-economic impact of the expansion of Norman Wells (Rees, 1986). It arose following an approach from UBC researchers concerned that vital elements of the impact of the pipeline development were not being sufficiently covered, and following a series of workshops and discussions led to community-based research in which local fieldworkers collected information for a comprehensive data-base in their communities. This, along with the comments of residents, could then form the basis for meaningful socio-economic planning which would allow people to combat the impacts of population increase and other non-aboriginal intrusions into their land. Other bonuses included research training for local staff. The ultimate goal was not only to collect information of use at both regional and community level but also to equip people more effectively to deal with similar demands and needs in the future.

The East Kimberley Impact Assessment Project was a multi-disciplinary research study conducted between 1985 and 1989 following requests from the Aboriginal people of the region. As Ross (1991) summarises, this remote region of Western Australia had since the 1950s suffered from severe pressures caused by a number of resource development projects in the region: the damming and subsequent flooding of the Ord River; the rise and fall of the pastoral industry, the tourist industry and development of Purnululu National Park and the exploration and development phases of the Argyle Diamond Mine. All of these had affected the Aboriginal population most of whom, by the early 1980s, were living in substandard conditions in small communities on the outskirts of towns such as Halls Creek, Kununurra and Wyndham. EKIAP, organised by a group of academics and funded from a number of academic and research institutions, was established to address these issues. The researchers worked very closely with the East Kimberley Aboriginal people, not only to ensure that the project reflected their perceptions of development, but also to equip them both to plan for the future and to devise strategies for implementing these plans. Although this element of the project was only partly successful, the academic researchers did succeed in recording vitally important information about the development history of the region. A number of workshops provided opportunities for exchanging ideas. At the first of these meetings over 400 people, largely from the area, but also including interested Aborigines from other parts of Australia and non-Aboriginal academics and government officials, spent several days exchanging views on a range of issues including mining, royalties, land use, tourism, parks, and employment and incomes. In general the project succeeded in raising Aboriginal awareness of external economic and political situations to a level where people were much more skilled at dealing with ongoing developments affecting them. As Ross (1991) states, some of the strategies which came out of this study have assisted other Aboriginal groups in dealing with similar problems.

6

THE ROLE OF PARKS AND TOURISM IN ABORIGINAL DEVELOPMENT

Remote parts of Australia and Canada have another resource, the value of which has only been recognised more recently. It is the land itself, increasingly attractive to tourists. Visitors are still overwhelmed by the emptiness of the deserts or the Arctic wastes; but they also notice its beauty; how the ochre landscape of the red centre, with its regular lines of parallel dunes, changes with the light; setting sun on the snows of the winter tundra; and the dappled light shining through the branches of fringing mangroves. Other things, like the silence and the solitude, also fire their imagination. Feelings like this have made the landscapes in these parts of Australia and Canada resources in themselves, places that people want to visit and explore. Some of them come on package tours, following a predictable round of journeys to key attractions and staying in hotels and resorts similar to those found in many other parts of the world; however increasing numbers are adventure tourists, part of the eco-tourism movement which promises to bring people into closer contact with both culture and nature. The need to escape from large cities like Melbourne, Toronto, Tokyo or London has become strong and the construction of all-weather highways and airstrips, of hotels, restaurants and souvenir shops have all helped to lure people into remote areas.

The growth of tourism has been welcomed by those concerned with remote area development. Tourism promises extra sources of revenue to bolster economies suffering from decline in other industries such as pastoralism or fur trapping; and, with the wealth of different demands brought by tourists, it has created an industry of many parts, with plenty of scope for innovation and ingenuity. However, despite intensive promotion, it does not yet dominate the economy in either the Canadian Arctic or the centre and north of Australia. In 1988, for example, it accounted for less than 1 per cent of total NWT revenue (75 per cent of which comes from federal government grants); in Australia's Northern Territory in 1989–90, dependent on federal money to approximately the same proportion, tourism, although more important than in Australia as a whole, only contributed 5.5 per cent of GDP. Thus, although its significance is undoubtedly growing, the hopes

189

Plate 6.1 Central Australian country: Aboriginal homeland and adventure tourist
mecca

of GNWT or the Northern Territory government that it will be the
foundation for future economic prosperity have yet to be realised.

The current interest in tourist development in these remote regions is
extremely important for aboriginal peoples who form very high proportions
of the local populations, who now control much of the land and whose
economic opportunities are otherwise limited. Their country, formerly seen
by non-aboriginal people as being of limited economic worth unless valuable
reserves of minerals could be proven, is now perceived to have other
attractions.

The relationship of aboriginal people to tourism in these regions is
different from that which they experience with any other form of resource
development. With tourism they themselves are part of the resource,
culturally and socially different from most other Australians and Canadians
and living an exotic lifestyle about which many tourists are curious. This has
created opportunities for them. However at the same time it has put them
under great pressure. In some cases aboriginal people have felt that the
intrusion in their lives, the loss of their privacy, is not worth the economic
gain to be made from taking part in tourism and have deliberately kept
visitors at bay. In others they have seen the opportunities offered by the
industry to be more important than these problems and therefore they have
chosen to join in.

THE TOURIST INDUSTRY AND ABORIGINAL PEOPLE IN REMOTE AREAS

In some ways the tourist industry in remote areas of Canada and Australia is not unlike the mining industry. It exploits a local resource, the environment, hopefully in a sustainable rather than destructive fashion; its large-scale activities, such as resorts, tour agencies, car hire firms and fast food and souvenir shop chains, are dominated by outside ownership, and a high proportion of the profits flow directly outside to other regions; tourist resorts also, like mining towns, are often developed in isolation from the region in which they are situated, with only limited linkage to that local region; and many of the people who work in these resorts are incomers rather than locals. Contrasts with the mining industry include the fact that the tourist industry must preserve the resource on which it depends, the environment; and that alongside its large-scale enterprises it has spawned many small-scale activities – tours and entertainments, arts and crafts, camping facilities – which offer much more varied opportunities for local involvement than usually come from mining. These differences have important implications for aboriginal peoples living in areas attractive to tourists. While many aspects of large-scale tourism are not compatible with aboriginal ways of life, smaller-scale activities can offer the chance to earn cash through doing jobs which use their own skills and which are enjoyable; and the need for tourism to be respectful of the environment accords with aboriginal perceptions of the importance of environmental husbandry.

The tourist industry brings with it both benefits and costs, in environmental, economic and social terms. Since tourism in remote areas is often closely associated with places of outstanding environmental interest and value, it has encouraged the designation of important areas as conservation zones, often as national parks, and has therefore been responsible for implementing strong legislation to preserve the environment. This, coupled with complementary legislation protecting places of spiritual significance, has on the whole benefited aboriginal people. Tourism can bring money: through direct employment in hotels, with tour agencies or in other activities; or through chances to sell locally manufactured goods, particularly arts and crafts. Investment in tourism – through becoming owners or part owners in resorts or other tourist ventures – can generate cash incomes and can also provide a useful basis for future development. Moreover tourism can be of cultural benefit. It can stimulate activities, such as making artefacts, painting, or conducting ceremonies, which were otherwise in danger of decline. This can be particularly important in ensuring that such information is passed on from older to younger people. Aboriginal people also value this type of activity because it helps non-aboriginal people to understand them in a more sympathetic and positive way. Educating the tourists in this way helps to break down prejudice and barriers which have proved to be major stumbling

191

blocks to appropriate forms of development.

None of these things comes without cost. Rising levels of income also mean rising consumerism and, as with the mining industry, growing social problems such as alcoholism and motor accidents. And environmental protection, even for well-resourced parks agencies, is not easy to enforce at all times and the sheer weight of tourists may cause environmental and social destruction which aboriginal peoples find quite unacceptable.

The key to successful aboriginal involvement in the tourist industry is control. Aboriginal people need to be able to exert at least some control over the basic resources which support tourism – the land and means, including financial means, to develop tourism on that land. This allows people to choose whether or not they want to be involved in tourism and what kind of involvement is acceptable to them; and to strike the correct balance – between preserving their social and cultural integrity and their privacy, and benefiting from the cash which they can generate through commercial development. The following descriptive scenarios reveal some of these important points.

Anangu in Uluru (Young, Fieldnotes, various visits, 1983–92)

Uluru-Kata Tjuta National Park in Australia's Red Centre – home of the national icon Ayers Rock (Uluru), now globally acclaimed as 'the world's largest pebble' – already receives over 250,000 visitors every year. As the tourists stop at the park entrance gate to pay their fees they see two signs. One presents a simple message: '*Pukulpa Pitjama Anangku Ngurakutu* – Welcome to Aboriginal Land'. The other: '*Nganana Tatinatja: Wiya – Anangu* Never Climb' is more subtle. It is shorthand for:

> Uluru is very sacred to us, and we would never think of climbing it. But we know that many of you, non-aboriginal people from all over Australia and from other parts of the world, have come here expecting to climb the Rock. When you climb it you remind us of termite ants going in and out of their nest. We will not stop you climbing. But we would like you to think about it first.

These simple messages demonstrate the finely tuned balance which underlies the management strategies of Uluru. This has been achieved through a delicate compromise between *anangu*, the Pitjantjatjara people who own the 'Rock' both in traditional terms and as Aboriginal freehold land under non-Aboriginal law; and the Australian National Parks and Wildlife Service (ANPWS) who run the park for *anangu* and the tourist industry.

At Uluru ranger station there is more evidence of *anangu* involvement in local tourism. An Aboriginal ranger sits at the information desk dealing with the never-ending flow of questions about climbing Uluru, driving to Kata Tjuta and, more mundane, the location of public toilets in the park. Outside the station a small group of tourists gather beside a 'witchetty bush' under the direction of two elderly *anangu* women who, with the help of a young Pitjantjatjara-speaking non-Aboriginal ranger, explain what lurks in the roots

192

of the tree. The tourists come from a mixture of national backgrounds – Japanese, German, American and Australian — and strain to understand the English translations of the information flowing from their guides. They set off slowly on the 'Liru Walk', a two-hour slow meander through the bush during which *anangu* will demonstrate and explain to the visitors many key socio-cultural and economic features of their lives. These include finding and collecting witchetty grubs, making fire without matches, working spinifex resin into a gum which can be used to anchor spear heads and axe heads in their shafts and relating the ancestral stories of Uluru – how the conflicts between *liru*, the poisonous snake, and *kuniya*, the python, led to the strange weathering patterns on the southern side of the massive outcrop.

At the rear of the ranger station visitors are crowded into the kiosk. This enterprise is owned by Mutitjulu, an Aboriginal community located about 4 km away inside the park. Many of the *anangu* at Mutitjulu are traditional owners of this country. They, along with relatives living in more distant Pitjantjatjara communities, form the core group which, following the handover of title to Uluru by the federal government in October 1985, provides the members of the park's Board of Management. The kiosk manager is non-Aboriginal and the store appears to operate like any other such outfit in Australia. However, stock carried by the kiosk bears witness to Aboriginal management policies – T-shirts, postcards, books and other souvenirs have all been carefully vetted and some have been specially designed in order to meet *anangu* perceptions of what tourists should be offered.

Maruku Arts and Crafts centre, next door, is a less conventional retail store. Six or eight traditional shelters made of bush materials (*wiltjas*) house a wide variety of artefacts: paintings in the western desert 'dot' and Pitjantjatjara styles; boomerangs, spears, coolamons (carrying dishes); and carvings of every conceivable kind of local native animal and reptile – snakes, lizards and, in every shape and form, goannas. Some introduced products – knitted bags made from imported wools – are also in evidence. In one area *anangu* women are carving, chattering amongst themselves and, on request, showing visitors their work; in another, the sales counter, *anangu* staff are completing credit card slips and wrapping up purchases for mailing directly to Japan.

A few kilometres from the ranger station lies the base of the climb. Here *anangu* are noticeable by their absence. Buses, trucks and cars stand in ranks, disgorging and collecting their passengers. The lines of 'ants' ascend and descend the bare face of Uluru, some moving fast and with confidence, others clinging to the chain or sliding down the hill on their bottoms. Many people do not go far up. Others, who have been all the way to the top and back, return determined to go straight to the gift shop in Yulara to buy a T-shirt which tells the world 'I conquered the Rock'. The Mutitjulu people, in line with their message about the climb, have banned the sale of this type of T-shirt from their kiosk.

Finally, as evening approaches, the tourists, heavily armed with cameras and tripods, descend upon 'Sunset Strip'. Bus after bus drives in and the people congregate on top of the sand dune along the fence, gazing at Uluru, exclaiming at the colours and arguing about the best time to 'snap'. Will it get even more spectacular? Or will the light begin to fade? Groups pose as an Uluru foreground, some rather incongruously clutching glasses of champagne and caviar covered biscuits. As conversation drops, camera clicks are clearly audible in the still air. With the dark they trek back to their buses, home to Yulara. Once again, *anangu* are absent.

193

Plate 6.2 Kara-Tjuta from the air

Plate 6.3 Descending from the Uluru 'climb'

Plate 6.4 Uluru and the hoards of buses at 'sunset strip'

This description tells us some important things not only about how tourists use Uluru but also about how the *anangu* want to be involved with the industry. They, as owners of the land, are of course able to exert a high level of control over what is provided, and many facilities in the park bear obvious signs of their influence. They do take on jobs, both in conventional 'shop-front' positions and, by their own choice, as part-time workers in less prominent activities; they have taken advantage of the opportunities for secondary enterprise development through the arts and crafts industry; both through their direct contact with visitors and through the marketing of their crafts, they deal with an international as well as national market; *anangu* try very hard to give tourists positive understanding of their society and culture; and *anangu* will rarely choose to be involved in any aspects of tourism which they find inappropriate, such as the climb or sunset strip.

Tourists in Canada's high Arctic (Young, Fieldnotes, 1984)

Cambridge Bay, on Canada's Arctic Victoria Island, is a large Inuit community, the regional centre of Kikitmeot district. Its people live partly off the land, spending varying periods of time in outcamps where they hunt game such as caribou and musk oxen and fish for the valuable and tasty Arctic char. They are also involved in the cash economy, drawing their incomes from working for organisations such as the council or their own co-operative or from social security. They are also expert at making soapstone and ivory carvings and clothing trimmed with exotic furs, valuable items for which there is little local market. Tourists very rarely venture as far as Cambridge Bay. When they do, their direct impact is quite obvious.

One early autumn morning in 1984 the ice-breaker anchored in the bay with the news that a cruise ship, with a hundred wealthy tourists on board, had just carefully nosed its way through the infamous northwest passage from the Atlantic and was close at hand. Immediately activity in the Ikaluktutiak Co-operative, owned and operated by the Inuit people, intensified. In the arts and crafts section, where stock had been built up following advance warning of this visit, they were eagerly anticipating a good sale. People unpacked and arranged carvings and artefacts and discussed prices, and two of the women agreed to demonstrate their carving techniques when the tourists arrived.

Within an hour of landing in the first chilly sleet storm of the winter, most of the tourists were crowded into the small store. They milled around the tables, quickly making inroads into the stock of small easily transportable items and also large soapstone carvings which had been awaiting buyers for months and even years. The shop resounded to discussions in Japanese, German and English from Texas, California and London. One particularly expensive carving, a soapstone figure of a musk ox priced at $3,500, was much admired and ultimately, as a number of people became seriously interested in acquiring it, caused some acrimonious arguments among the tourist group. Even as the Texans were signing the cheque the Germans were still loudly claiming that they had seen it first. One Inuit privately commented that perhaps the musk ox might have to be cut in half, but on a more serious note observed regretfully that if they had been given more accurate information about these tourists they could have made more than one musk ox. Finally, at the end of the morning, the tourists departed to the ship, deciding that a walk around Cambridge Bay was an unattractive prospect in sleet and a howling gale. The Co-op staff spent the afternoon packing carvings and other items for the mail plane, and next morning the cruise ship departed on the rest of its journey to Japan. Life at Cambridge Bay returned, at least on the surface, to normal.

This visit illustrates important factors affecting the efficient operation of the arts and crafts industry, a very important complement to tourism in remote areas. It provided quite a windfall to the Co-op: sales of valuable stock unlikely ever to be purchased locally; selling directly, without having to incur extra costs caused by freighting goods to southern outlets; and inevitably increasing outside information of what the skilled craftworkers and artists of Cambridge Bay were able to produce. But it also highlighted characteristics which are an inherent part of aboriginal involvement in the tourist industry in the high Arctic: the problems of remoteness, of communications, of the small scale of local operations and their lack of potential to cope with sudden changes in demand and supply for products. Another feature was the curiosity which the tourists displayed about the Inuit, continually asking questions and making loud observations, many of which would certainly have been overheard and understood. Although this was not on the whole obtrusive it still created some pressure. Counteracting this was the high level of control which the Inuit, both Co-op workers and artists, exerted over the whole operation.

Plate 6.5 Cambridge Bay Inuit Co-operative store

The tourist experience in town (Young, Fieldnotes, various visits, 1983–90)

For international and many interstate visitors Alice Springs, the tourist mecca in Australia's Red Centre, first appears from the air as a sprawling town squeezed between the spectacular escarpments of the Macdonnell Ranges and merging well with the red ochre landscape. From the airport visitors are swiftly transported the 20 km along the bitumen highway, through Heavitree Gap where the Todd River has incised itself through the hills, to their hotels – high-quality resorts which are operated by large-scale international and national companies and where every modern convenience is available. Indeed, for many purposes there is little need to leave the hotel – food, shops, swimming pools and spas and even a casino are all there on site. Only for the viewing of Alice Springs and its surroundings, surely the main reason for coming here, is further travel essential. Air-conditioned coaches, provided through arrangements made by the hotel, make this possible and the tourists take off for the standard tour – ANZAC hill, where a panoramic view of the town is presented, the Old Telegraph Station, the original reason for non-Aboriginal settlement in the area, and Todd Street, the original main street whose weatherboard verandahed buildings have now been replaced by a post-modernistic shopping mall like many found elsewhere in Australia and overseas. On the way they cross and recross the dry bed of the Todd River, home of the famous (or infamous) Henley-on-Todd regatta in which bottomless boats are raced through the sand while the grog flows very freely.

As at least some of the tourists realise, this offers a very limited view of Alice Springs. A common question from the more sensitive and inquiring souls is:

> Where are the Aborigines? We came here to see them. We expected them to be working in the hotels, or acting as our tour guides. But all we have seen are their stunning paintings and artefacts in the shops, and many of these shops are run by Europeans. Why are they not taking centre stage in the industry?

The answer, in many cases, would be:

> They're too lazy to work. If you want to see them look under the trees in the Todd River where they'll be sitting with the grog bottle. But don't go there at night.

Such a scene may be somewhat exaggerated because today there has been increasing Aboriginal participation in town-based tourism. Nevertheless it does highlight the main features of the situation – that tourism in these locations is big business, and that much of the industry is not locally controlled, let alone Aboriginally controlled. Aboriginal participation is restricted by these characteristics. It is also restricted by the alien nature of the operation, and by Aboriginal reluctance to put themselves into promi-nence in an industry which demands direct contact with a general public who may well be aggressively ill-informed, critical and racist in their attitudes. It is not representative only of Alice Springs. In essence such scenarios also

197

occur in Inuvik and Yellowknife and Inuit, Indian and Aboriginal people would find many points of agreement in discussing how tourism affects them.

Aboriginal involvement in remote area tourism has become a significant development issue simply because the regions where they live are increasingly attractive to the present-day tourist. Prime considerations are whether they own the areas used by tourists, whether they control the development of facilities within those areas and have the financial means to carry out whatever forms of development they favour, and their attitudes to the social pressures imposed on them by the industry itself. These considerations are all affected by a combination of environmental, economic and social concerns.

ENVIRONMENTAL CONCERNS

Remote parts of Canada and Australia obviously differ in their attraction for tourists. For decades, Australian locations such as Queensland's Great Barrier Reef, Uluru or Alice Springs have drawn tourists because of their scenery, environmental diversity and historical associations. Other places – Kakadu, the North Queensland rainforests, the beaches at Broome and Purnululu (Bungle Bungle) – have joined the list more recently. Some of these locations lie on or near Aboriginal land. They are the pressure points, well established on every basic tourist list, and the places where conflicting environmental issues are most likely to occur. However vast areas of Aboriginal land in remote Australia offer little to the conventional tourist. Most people find the huge stretches of flat and dune covered country in the Tanami or Great Victoria deserts too intimidating and alien to come to grips with, and the sparseness of life, both human and animal, on that land adds to its forbidding nature. It is only within recent times that this isolation and emptiness has itself become an attraction for less conventional, adventure tourists. While the activities of these people generally have a much smaller impact, there are still environmental issues to be considered.

Canada's experiences with remote area tourism have, on the whole, been more limited. Until very recently the harsh environments of the north have precluded all but small-scale development and as a result the direct impact of this industry on Indian and Inuit communities has been quite limited. However growing awareness of the unique beauty of the Arctic and of its wildlife, coupled with improvements in communication systems to the area have led to increased tourist interest. Some aboriginal groups are therefore now coming face to face with tourists on their doorsteps. Recognition of the fragility and vulnerability of northern Canada to other forms of development such as mining have led since the 1970s to the establishment of some National Park Reserves, such as North Yukon, Kluane and Auyuittuq.

Tourism in both of these regions affects both towns and rural areas. In the towns, where most of the conventional motels and recreational facilities such as shops, tour companies or casinos are situated, the interaction between aboriginal people and tourism may be quite limited. These places house a wealth of different tourist operators and the complexity of the industry, particularly as it relates to the provision of services, is great. Tourists themselves tend to be more demanding in terms of facilities – they will put up with 'roughing it' in the bush but in town they expect things to operate much as they would in their home towns. The general outcome is that aboriginal people, although they form a high proportion of the population, may, as the brief Alice Springs scenario presented above suggests, have little direct involvement in the industry. It is in the rural areas that the environmental issues affecting remote area tourism and aboriginal people come to a head. Many of these issues focus on the development of national parks and other forms of reserve.

Parks, aboriginal peoples and their land interests

National parks and the tourist industry often work in tandem. The establishment of a park may well encourage the development of tourism, particularly if it is accompanied by the development of facilities; and, alternatively, uncontrolled tourist use of an area may lead to it being designated as a national park, in order that it can be protected. The designation of significant parts of the Canadian and Australian north as national parks reflects a number of factors: realisation of the threats posed to such areas by forms of resource development such as mining, coupled with knowledge of the unique nature of their habitats for both wildlife and plant species; perception that these areas have little alternative value and therefore might usefully be zoned as conservation regions; and increasing pressure from urban-based populations who, with their growing environmental knowledge and interests, feel strongly that conservation of these regions is essential.

National parks have been conventionally defined, according to the International Union for the Conservation of Nature and Natural Resources (IUCN,1970), as areas where 'one or several ecosystems are not materially altered by human exploitation and occupation' and 'where the highest competent authority of the country has taken steps to prevent or to eliminate as soon as possible exploitation or occupation in the whole area'. This definition, stemming, as Stevens (1986) points out, from the Yellowstone model in the United States, stresses the value of a park as a wilderness area and explicitly precludes human occupation within its boundaries. It also anticipates that evidence of former human use will be eliminated. It obviously presents difficulties for parks on aboriginal-owned land, where traditional owners not only want to live within the park but also want to be able to continue harvesting subsistence resources and carrying out traditional land management practices such as controlled burning. This highlights important issues over the designation of areas of aboriginal-controlled land as national parks, issues which may well place them in conflict with park authorities. Indeed, as Head (1990) discusses, there are inherent dangers in assuming that the views of aboriginal people and conservationists on the prime roles and functions of parks will coincide. These possible areas of conflict will affect aboriginal input and gain from the tourist industry. Australia and Canada provide some interesting examples of how these issues are being dealt with.

Parks and Aboriginal land interests in remote Australia

In Australia, significant areas of Aboriginal-owned land, principally in the Northern Territory, have been set aside as national parks and wildlife reserves (Figure 6.1). Their Aboriginal owners have accepted this strategy with a certain reluctance. In some cases they have had no choice. With Uluru, Kakadu and Nitmiluk, the designation of the land as national park was a condition for the granting of the land claim – no park, no land. However subsequently, both there and elsewhere, people have realised that there are some environmental and cultural advantages to be gained from having land declared as parks and have accepted this approach. Tourism has, almost inevitably, followed in its wake.

The longest established and best known Aboriginal-owned national parks are in Australia's Northern Territory: Kakadu, declared in 1979; Gurig (Cobourg), in 1981; Uluru/Kata Tjuta, 1985; Nitmiluk (Katherine Gorge), 1989 (Figure 6.1). Kakadu and Uluru, declared as parts of Aboriginal land trusts under the Aboriginal Land Rights Act, (NT), 1976, were leased back to the Director of the Commonwealth Australian National Parks and Wildlife Services (ANPWS) for management and development. Nitmiluk, also declared as Aboriginal land under the Aboriginal Land Rights Act and Gurig, established under the Cobourg Land Act of 1981, are managed by agreement with the Conservation Commission of the Northern Territory (CCNT) and the traditional Aboriginal owners, represented by the Northern Land Council (NLC). Each of these parks, as a number of existing studies (for example, Weaver, 1984; Altman, 1988) have commented, operates under its own distinct management plan. Their plans reveal some important common themes – a commitment to Aboriginal involvement, through day-to-day management including employment in the park ranger service; special provision of cultural interpretation; the development of tourist-oriented enterprises in the parks; and planning the future development of the park itself. Parks have, for these particular Aboriginal groups, become both a very important form of land management and a significant avenue for development.

Aboriginal interests in access to parks and direct participation in park

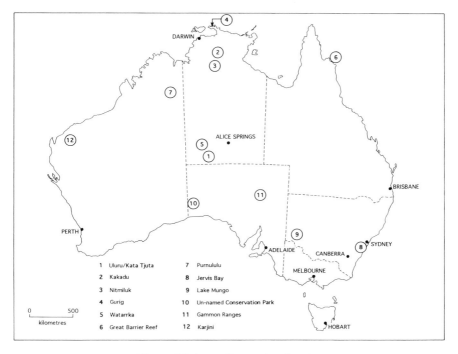

DARWIN

ALICE SPRINGS

BRISBANE

PERTH

ADELAIDE   CANBERRA

SYDNEY

MELBOURNE

HOBART

1	Uluru/Kata Tjuta	7	Purnululu
2	Kakadu	8	Jervis Bay
3	Nitmiluk	9	Lake Mungo
4	Gurig	10	Un-named Conservation Park
5	Watarrka	11	Gammon Ranges
6	Great Barrier Reef	12	Karjini

0   500
kilometres

Figure 6.1 Australian national parks

management and the tourist industry are not, however, confined to those areas to which they legally hold title. Their traditional ownership responsibilities extend into areas held by other lessees, including parks authorities. Such a situation certainly exists in the Northern Territory where many parks have been established on land leased by the parks authorities. Elsewhere it is the dominant situation, because Aborigines lack the legal backing to push for control over park management policies. Nevertheless, although they cannot exert that power, their claims to be involved in park development are increasingly being regarded with sympathy and state governments are beginning to draft agreements which give Aboriginal traditional owners a significant input (Gillespie, 1990). These have included joint management of some parks. Parks where such discussions have already occurred include the Un-named Conservation Park (covering parts of the Great Victoria Desert and designated in 1970), Gammon Ranges National Park and the Nullarbor National Park and Regional Reserve (all in South Australia); Purnululu (Bungle Bungle), in the East Kimberley and Karjini in the Hamersley Range (Western Australia); Mootwingee, Lake Mungo and Jervis Bay National Parks (New South Wales); and Watarrka and Macdonnell Ranges National

Parks in the Northern Territory. The Great Barrier Reef Marine Park in Queensland has so far been omitted from these discussions. This is a major concern to neighbouring communities such as Yarrabah and Palm Island because there has been severe tourist pressure on the reef for many years. They have already taken the initiative by monitoring the resource harvesting activities both of their own people and of outsiders on the reef area.

National parks and aboriginal interests in Canada

Parks Canada, the federal body responsible for Canadian national parks, has maintained its determination to retain title to the land concerned, and thus, as Gardner and Nelson (1981) point out, the kind of leaseback arrangement which has occurred in Kakadu and Uluru is not possible. This is unfortunate because, as Inuit and Indians have clearly indicated, they are very concerned about the possible conflict between the use of park lands and their resources for recreation and for subsistence.They want as high a degree of local aboriginal control as possible. Over recent decades Parks Canada has increasingly recognised aboriginal interests in hunting, in resource management and in the control of tourist activities. By 1979 they had also begun to discuss the possible creation of joint management systems. However Weaver (1984) noted that, five years later, no such arrangements had yet been finalised. Land claim negotiations have brought these issues into greater prominence. During the many years of discussion over the terms of claims, Parks Canada and northern aboriginal peoples agreed that no areas within the regions covered by comprehensive land claim legislation would be declared as national parks until the claims themselves were settled. In the interim such conservation areas in northern Canada remained as national park reserves, a designation which allowed for much greater flexibility and the opportunity to work out more acceptable compromises within the final agreements. Parks have subsequently been declared in conjunction with the COPE claim and, much more recently, as part of the Nunavut settlement. Fenge (1993) has recently stated that the Inuit have seen the declaration of parks in Nunavut as an important way of increasing the land which comes under aboriginal control. When land was declared as park it was excised from the claim and hence the Inuit share, on a proportionate basis, was calculated from what was left. This has meant that more land will receive the kind of environmental protection which the Inuit want.

The extent to which Canadian aboriginal peoples can influence development and management policy in northern parks is very varied. In Yukon Territory, for example, aboriginal involvement in Kluane, which has not been associated with any land claim negotiations, has generally been discouraged and they have no influence on its management; however North Yukon, established in 1984 under the final agreement for the COPE land claim, was the first park to include recognition of aboriginal hunting, fishing and

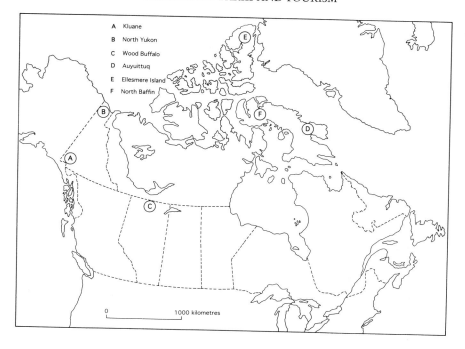

Figure 6.2 Canadian national parks

trapping rights within its boundaries and also to pave the way for a joint management agreement (Figure 6.2). Similar arrangements for Wood Buffalo and an area on the east arm of Great Slave Lake straddling the NWT and Alberta/Saskatchewan borders, were discussed as part of the Dene/Metis land claim negotiations but these talks have now ceased with the collapse of that process. However the Inuit have ensured that the recently signed Nunavut agreement includes provision for full recognition of Auyuittuq, Ellesmere and North Baffin Island as parks. Their attempts to include two additional parks in this agreement – Bluenose (near Coppermine) and Wager Bay (on the western coast of Hudson's Bay – have not been successful, principally, Fenge (1993) thinks, because Parks Canada lacks the resources to take on additional commitments in the north. Fenge also comments that the Inuit have recently been indicating that they consider the economic development role of parks to be increasingly important (see p. 273).

Parks as an aboriginal development strategy

Aboriginal people have often been suspicious about the real reasons why government agencies have pushed for the declaration of national parks in remote parts of Canada or Australia. They have seen it primarily as a method of making land ineligible for claim under land rights legislation. These beliefs have been borne out in Australia by submissions and evidence presented by park authorities at land claim hearings, where there have been some vigorous attempts to excise areas prior to the claim on the basis of their anticipated conservation value. Similar pressures have been exerted in Canada. However, after resolution of these conflicts, park development has often appeared to be a more attractive option. The environmental advantages of this, particularly where there is strong aboriginal influence on park management, can be considerable.

From an aboriginal perspective, park declaration can help to provide environmental assistance, looking after the country, by eradicating feral species of plants and animals in favour of the regeneration of native species; and through protecting places of cultural value. In Australia parks authorities have supported programs to eradicate feral animal species such as rabbits, goats and cats. Complementary programs to reintroduce native species affected by feral animal depredations have also been set up. Recent examples include the reintroduction of the rabbit-eared bandicoot, the giant bilby, into Uluru. This animal, although still surviving in more remote parts of the desert, had disappeared from the Uluru region following territorial competition with rabbits. Another much more highly endangered species, the rufous hare wallaby, the *mala*, is being reintroduced to the Tanami desert following a breeding scheme run by the Northern Territory's Conservation Commission in Alice Springs. Aboriginal feelings about these programs have been generally supportive although there has been concern over the eradication of rabbits because they now form an important 'bush tucker' item. They have been particularly enthusiastic about the *mala* scheme. This animal is of great cultural significance to many Central Australian Aborigines and there has been widespread concern over its disappearance. The level of enthusiasm has even been sufficiently great for people to resume the challenge after a disastrous first attempt at reintroduction which resulted in the decimation of the *mala* in less than a week through the activities of one particularly determined feral cat.

Park authorities have also, after some initial scepticism, used some Aboriginal land management practices to enhance the regeneration of natural vegetation and wildlife habitats. This has included controlled burning to inhibit the spread of spinifex and kill introduced plants. These techniques have depended heavily on Aboriginal advice, and have therefore given Aboriginal people practical input into park management in both Kakadu (Press, 1987) and Uluru (Saxon, 1984). The result, as can now be seen at

Uluru, is a landscape which approximates more closely that of earlier times, before the depredations of hard-hooved animals such as cattle took their toll.

Protection of sacred sites is another positive contribution of national parks. At Uluru and Kakadu the restriction of tourists to only a few 'sacrificial sites' has had a very positive effect in keeping tourists away from other more sensitive places. *Anangu* have commented that when they manage tourists in this way it is like mustering cattle; you organise them into groups and then make them go where you want them to go,

Where aboriginal people are unable to influence management policy park development can have negative environmental consequences. Comparison of the 'before' and 'after' situation at Uluru is instructive. Prior to the handover of Uluru to Aborigines in 1985 entry to the park and activities around the 'Rock' were much less restricted than they are today. People wandered at will around the base of the monolith, in and out of rock shelters where they desecrated paintings, and over country which was environmentally sensitive. Special events, including rock concerts held beside, or even on top of the Rock, and visits by car rallies, were uncontrolled, and film companies could record at will. Such activities caused great concern to *anangu*, but they were unable to do anything about it. Uluru today is quite different, a compromise between what the tourists want to do and what *anangu* perceive as appropriate; the Rock is still climbed, some caves are still visited but many other places are protected. Although some of today's tourists complain about these new restrictions many understand and are sympathetic. In its increasingly important role in educating tourists about Aboriginal cultural and environmental practices, the park is making a significant contribution to cross-cultural understanding.

ECONOMIC ISSUES

Aboriginal people derive economic benefit from tourism through employment. However the extent of aboriginal employment in tourist enterprises is, from observation, limited. In general aboriginal people lack the educational qualifications and experience to find jobs with hotels or tour companies unless these employers are specifically seeking people with cultural knowledge which can add an authentic touch. Moreover, the type of work involved in hotels is not avidly sought by most aboriginal people, not only because of their inexperience but also because these jobs continually thrust them into direct contact with non-aboriginal tourists. Many tourists are friendly and interested in learning about aboriginal people; however this puts aboriginal workers strongly into the limelight, in ways which they often find quite stressful. Other economic benefits coming from the tourist industry occur through aboriginal ownership and/or investment in tourist enterprises; and, less directly, because of improvements in infrastructure such as roads, services or housing, all stemming from the demands of tourist development.

Employment

Park rangers

Aboriginal people have shown a preference for jobs, such as park ranger positions, where they can make an obvious contribution because of their particular skills and knowledge. Moreover, many of their mutual agreements on park management have included clauses giving them preference in these types of jobs.

In Australia opportunities for Aborigines to work as park rangers have increased markedly within recent years, and many Aborigines are finding the job highly satisfying. Appropriate training for these positions has become an important issue. Early Aboriginal ranger employment schemes, largely in the Northern Territory, brought in traditional owners of the land within the park, largely in jobs which relied on their knowledge of the local environment and in which they could make a distinct cultural contribution. Many of these earlier rangers were older people. However more recently, as younger Aboriginal people with higher levels of formal education have come into this work, there have been deliberate moves to give them experience in all elements of a ranger's job, including taking charge. This approach will obviously help to increase Aboriginal influence in tourist development. It has meant a demand for more sophisticated training, including the creation of a proper career path for Aboriginal rangers. Current courses range from those leading to a bachelor's degree, diploma courses specifically aimed at working for federal and state/territory parks services, and other courses which are suitable not only for working in the parks services but also for people wanting to work as rangers within their own communities. This last initiative, which started in answer to requests from Aborigines in the Cape York area of Queensland, is an interesting spin-off from the tourist industry. It is helping community groups to monitor resource use in areas where pressure is coming both from their own growing populations and also from visitors. Regardless of the level of course taken, all trainee rangers learn most of the job through intensive on-the-job experience.

These training schemes have had a positive although somewhat patchy impact on Aboriginal ranger employment. In mid-1990 one-third of Kakadu's rangers, thirteen people in all, were local Gagadju people (Young *et al.*, 1991). Several had worked at the park for a considerable time and they varied widely in age and skills. Some were younger people with relatively high levels of education; others were older people, community leaders with extensive traditional knowledge of the area but lacking formal education. The Gagadju people valued this work highly because they felt that it allowed them to present important role models to others in the community, as well as ensuring that there was a strong day-to-day Aboriginal input into the management of the park. *Anangu* at Uluru have felt differently about it. In

1990 that park had only one full-time Aboriginal ranger. This did not mean that *anangu* were less interested in looking after the park. On the contrary, they have played a very important role in this. However they seem to feel that they can contribute best in decision-making rather than in full-time direct contact jobs. This partly reflects their more limited education and experience of working with non-Aborigines. As a result those who do work with the parks service largely do so on a casual basis, sharing the job of conducting interpretative tours for small groups.

Outside the Northern Territory Aboriginal ranger employment has also become much more firmly established. South Australia, for example, introduced a ranger training program for the Gammon Ranges National Park in the early 1980s and has subsequently extended this to other parks. Interest in park ranger training has also been expressed in Western Australia but CALM has not yet been able to establish a scheme because of current lack of funding. Some trainee rangers from that state currently attend the courses offered in the Northern Territory.

Inuit and Indian people have also shown interest in working as rangers, although to date the opportunities seem to have been more limited. This reflects the later establishment of parks in NWT and Yukon, and also the fact that those parks which do exist are much less accessible than their Australian counterparts. As a result tourist numbers are lower than those in places like Kakadu and Uluru and there are fewer employment opportunities. However programs to train aboriginal rangers for Kluane were already under way in the late 1970s (Gardner and Nelson, 1981: 211) and the agreement to establish North Yukon National Park included assurances of high levels of employment for the Inuvialuit claimants. By 1987/8, only four years after the claim was granted, 77 per cent of all park positions based in Inuvik were held by Inuvialuit (DIAND, 1988b: 12) and training programs were already being established. The principles and aims behind their involvement appear to be very similar to those which have drawn Australian Aborigines into this type of job.

Parks do not only employ people as full-time rangers or as casual guides. They also need part-time workers, often to carry out short-term contracts to improve facilities, create interpretative material and deal with day-to-day resource management problems. ANPWS in Australia has been running an Aboriginal contract employment scheme of this type for the last four years. In 1990/1 that program, with funds of $1.6 million, supported eighty-one conservation projects with Aboriginal groups throughout the country. Many of these concerned work in parks and were therefore directly concerned with the support of the tourist industry. These included interpretative projects, such as developing an inventory of Aboriginal placenames and documenting Aboriginal history; and construction projects, such as building campgrounds, tourist walkways and fences (Young *et al.*, 1991). In 1992, following recommendations from the RCIADIC for support for Aboriginal employment programs within communities, this program received a large boost in funding.

Interpretative services (Young, Fieldnotes, 1990–2)

Aboriginal people are, of course, the most appropriate group to show visitors around their country. However it has only been relatively recently that some communities have begun to set up their own interpretative businesses. That established by the Wreck Bay people from the south coast of New South Wales provides a particularly good example of the challenges and pitfalls which occur. Although this is not truly a remote community their experiences would certainly mirror those that others in more isolated parts of Australia are beginning to face (Young *et al.*, 1991).

The Wreck Bay community of about 150 people, on only 400 hectares of Aboriginal-owned land, is surrounded by Jervis Bay territory, an area which in the 1920s was excised from the state for the development of coastal access for the new inland capital of Canberra. That never happened. However the bay and peninsula, with beautiful beaches and relatively untouched bushland, have become an extremely popular tourist area and during the summer months receive a large number of visitors. Although the region has long been designated as a public recreation reserve it has only recently become a national park and the Wreck Bay people, along with their compatriots from another nearby community, have for some time been involved in negotiations over some form of joint management or at least strong Aboriginal input into park plans and policy. In the interim they decided to get ahead by establishing their own interpretative tours company.

Walkabout Tours started its operations at Wreck Bay in 1988 and has functioned, with somewhat mixed fortunes, ever since. Initially the enterprise was small and rather haphazard, employing only two people on a part-time basis and providing tours on demand from tourist groups mostly staying at the nearby camp ground. However since 1990 the business has expanded and become much more professional. Four different types of tour are now offered – bush tucker walks and marine walks on the wave-cut platform by day; and spotlighting walks and campfire meetings with story-telling about local Aboriginal history in the evenings. The guides, now including both men and women and older and younger members of the community, are kept extremely busy working with casual visitors during summer and, through a well organised publicity campaign, have built up a steady flow of work from school and university groups at other times of year. Their growing skills in the enterprise have led to requests for assistance from other coastal Aboriginal communities in the area, and they have passed on advice on how best to conduct this type of business.

Walkabout Tours started as an Aboriginal initiative – a deliberate effort to share in the economic benefits offered by tourism. High levels of unemployment in this community, where the fishing industry has declined to nothing, were a major incentive. However the business has not yet been economically independent. From the beginning external funding was necessary, and problems in obtaining that support caused many of the early difficulties. But since 1990 Wreck Bay has succeeded in tapping into funds from three federal government departments: ANPWS (wages for the two experienced guides, $50,000); ATSIC, ($20,000 for capital works, primarily the demountable office building); and DEET (training officer and funding for training additional guides, at least $100,000). Moreover these departments have managed to co-ordinate their input, an approach which has been all too infrequent in Aboriginal development support. In 1991 the group estimated that the business could bring in over $60,000 in its first year of operation. Although it is hoped that the enterprise will eventually become economically viable, it will probably continue to need government funding in the short term.

Why has Wreck Bay Walkabouts been successful so far? It is in many ways a classic example of how Aboriginal tourist enterprises can get off the ground. Firstly, it shows that community support and determination bring results, in terms of external support. In general terms the Walkabout Tours experience provides an interesting example of the 'next economy'. Secondly, the enterprise not only provides employment and training for a significant number of people in a community with high rates of unemployment, but is also a source of considerable pride to the whole group. They clearly demonstrated that they are able to start something up on their own. Thirdly, with the increasing pressure on resources in the Jervis Bay area, Wreck Bay people now have a much better chance of active participation in the new National Park and in the services which it offers. Finally the methods of support used for the enterprise demonstrate that co-ordination between funding bodies can and does occur.

Successful interpretative services can also be established by non-Aboriginal companies knowledgeable and sympathetic towards aboriginal cultural concerns. In such cases aboriginal people are usually willing to take on employment. Desert Tracks is one such Australian company. Since 1988 this organisation has been providing a unique tourist experience in the Anangu Pitjantjatjaruku (AP) lands in the northern part of South Australia. Their activities include camping with the local Aboriginal groups and learning at first hand about many cultural and economic facets of their lives. Although initiated, owned and managed by non-Aboriginal people it has always involved intensive consultation and cooperation with the Pitjantjatjara people and has benefited them in a number of ways. This has included providing them with a source of cash income from a type of work which they find congenial and for which they alone hold the appropriate skills. Social and cultural benefits include educating both younger *anangu* and also non-Aboriginal people about these matters, and getting people out on the land where they can more effectively take full responsibility for resource management (Snowdon *et al.*, 1990). As with Wreck Bay Walkabouts a key to the success of this enterprise is the small scale of its operation, which makes the whole experience manageable and not too intrusive for the Aborigines concerned. With non-Aboriginal ownership, the level of Aboriginal control has obviously been less. However, in compensation, the business has been run without government assistance and/or intervention. If, as has been suggested, Desert Tracks is transferred to Aboriginal ownership then these circumstances would probably change.

Aboriginal peoples and the hotel trade

The increase in tourist numbers, coupled with changes in the types of tourists has had a significant impact on accommodation development in remote regions. Only a couple of decades ago the bulk of tourists travelling in outback Australia came overland, and most anticipated either staying in basic motels or camping. Today not only main towns such as Alice Springs, but also much smaller centres like Broome and resorts like Yulara are on the international tourist network, and much of their trade is taken from visitors on package tours, increasingly from overseas. Similar trends have occurred in the Canadian Arctic, albeit on a smaller scale because of more limited access and greater cost. These tourists have different needs from their predecessors – large, modern hotels, often in resort developments; and camping facilities which are becoming more and more sophisticated. In 1990 the NWT could collectively provide accommodation for almost 3,500 persons in hotels and motels and additional space was available in lodges and camps operated by tour companies (Hamley, 1991: 393). While most of these hotels are non-aboriginal owned and are operated under contract to national or international companies, aboriginal involvement in this part of the industry has begun to increase. Capital resources are obviously a key factor.

Australian Aboriginal participation in larger hotel ventures occurs either through outright ownership or under joint venture agreements. Because of their strategy of investing their uranium royalty monies in tourist facilities, the Gagadju people of western Arnhem Land now own all three hotels serving Kakadu National Park. These are all managed under contract by non-Aboriginal operators including, in the case of the Jabiru 'Crocodile Hotel', a major national company. Ownership of the hotels has been a worthwhile investment for the Gagadju and it also gives them some control over tourism development associated with the park. It has not, as yet, brought them many jobs. This partly reflects their own choice, because of the strains involved in the nature of the work; it also reflects their limited experience in this type of enterprise.

The King's Canyon Resort development, opened in 1991 adjacent to Watarrka National Park in central Australia's Macdonnell Ranges, appears on the surface to be similar. However here the resort is a joint venture, owned by a consortium including ATSIC Commercial Development Corporation, Centrecorp (a company with shareholders from three of the main Alice Springs-based Aboriginal organisations) and a large NT tourist development company. Once again, very few local Aboriginal people have shown interest in working in the accommodation or conventional service aspects of the resort, serving in the restaurant or at hotel reception. However clauses within the joint venture agreement included encouragement of small businesses providing tours concerned with local Aboriginal cultural issues. These are already in operation. People from the traditional land-owning groups of Watarrka, several of whom are living in small communities within the park itself, enjoy working with these small, easily controlled tourist groups in tasks where they can use their unique skills and knowledge. With the probable expansion of the resort following road access improvements, this type of enterprise is likely to become more and more successful.

Not all visitors to remote regions are tourists. Hamley (1991: 395) notes that 42 per cent of visitors to NWT in 1989 were there for business purposes. They are mostly southern Canadians, bureaucrats spending short periods in the north from Ottawa and other cities, and contractors involved in construction projects run by southern companies. Although much of their work lies in towns like Whitehorse, Yellowknife and Inuvik, many also spend time in more remote, largely aboriginal communities. Here they stay in small community hotels. Most of these are aboriginal owned, often as one of the businesses run by the local co-operative. Cambridge Bay Hotel, where I stayed while visiting in 1984, is owned by Ikaluktutiak Inuit Co-operative. During the short northern summer, when most visitors come north, its accommodation for forty people is often fully used; however it is still quite busy in winter because groups of construction workers, who started their jobs in late summer/early winter when the ice break-up had allowed the barge to bring in building materials by sea, remained for six months. Most of the workers in this hotel are Inuit, although it periodically brought in a non-aboriginal manager. Although a small venture, its operation was fairly complex, with the major problems reflecting the high costs of essential services – power, water and sewerage are all prohibitively expensive and hard to maintain in a permafrost region such as this. As a component of the local co-operative the hotel played a central role for other Inuit enterprises such as the taxi service, the bakery, the arts and crafts store and the fish (arctic char) freezing plant because its visitors provided much of their trade. Hotels such as that at Cambridge Bay are scattered throughout northern Canadian remote communities and, while much of their business has not yet been in the tourist market, they will undoubtedly be able to tap into this trade if the demand occurs.

Most aboriginal people working in the tourist industry, whether for hotels or in service activities such as tours, comment that they prefer to be employed with aboriginal-owned companies. Here they are likely to find a more congenial mode of operation and sympathetic work companions. A popular alternative, usually providing even greater independence is self-employment in related activities, particularly in the arts and crafts industry.

The arts and crafts industry

The tourist industry has been a prime cause for the development of arts and crafts enterprises in aboriginal communities. For many reasons these enterprises are well suited to assist development in remote areas. Unlike most other tourist related ventures, this business is based on traditional skills which are an integral part of aboriginal culture. That, along with dependence of arts and crafts on locally obtained raw materials, the relatively high value of the products, which make the business cost-effective even in extremely

Plate 6.6 Women collecting materials for artefact manufacture in the 'Top End'
rainforest: tree-bark as a medium for paintings; saplings for spear hafts; vines for
making 'bush string'

remote communities and the nature of the work itself, which is flexible and
can be fitted in with other commitments, has favoured its development. In
both Canada and Australia the arts and crafts business has now become very
important for aboriginal people. With increased contact with international
markets it has also become more and more complex and also more likely to
be affected by broader economic issues and events well outside aboriginal
control. This has introduced a degree of vulnerability (Altman, 1989). The
protracted Australian airline pilots' strike of 1989, for example, caused a
decrease in tourism which in turn depressed arts and crafts sales; and the
Inuit arts and crafts industry has been badly hit by bans on sealing and on
use of ivory, imposed by important purchasing countries such as the United
States and Germany.

The following comparative analysis of the aboriginal arts and crafts
industries in Canada and Australia discusses several topics. These include: the
production process; the marketing of the products and who the customers
are; and a wealth of issues such as coping with the mix of economic and
cultural elements in production; the monetary return to artists and crafts-
persons; and the overall contribution of the industry to development.

The actual product is affected by differences in the types of customers for
whom it is intended. An important distinction, as Altman (1990) outlines for

Aboriginal art in Australia and Isaacs (1988) for the Canadian Inuit art industry, is that between 'fine art' and 'tourist art'. In the former case the product belongs within the conventional art world and, as Isaacs stresses, should be considered more as part of the mainstream of this world rather than in the 'ethnic' art stream to which it is often consigned. Artists working in fine art are often individually recognised by name, their products are frequently sold by exhibition rather than in conventional art stores and these are also, if of sufficiently high quality, on display in museums and art galleries. Prices for these pieces are relatively high and many customers purchase for investment as well as because they are discerning lovers of aboriginal art. Tourist art, on the other hand, is more mass-produced although it is still often hand-made. It is cheaper and items are often smaller and more portable. Small souvenirs are a main component of tourist art. Although the distinction between fine art and tourist art is important the two categories do often overlap and hence, as Altman (1990) points out, it is difficult to assess how much of the industry falls into each segment.

Arts and crafts products are extremely varied: paintings, spears, boomerangs, didgeridoos, carved wooden figures and implements dominate present day 'traditional' production in Australia; while in Canada decorative clothing made from skins and furs, jewellery and soapstone carvings are important traditional products. Other products derived from these bases – the Inuit drawings and fine prints which come from many Arctic communities; the T-shirts and materials with 'Aboriginal' designs; and other unusual inventions such as fridge magnets with mounted miniature carved wooden goannas, or giant mosquitoes with sealskin wings and leather legs and proboscies – are also featured in the modern industry in both countries. Characteristics shared by Canadian and Australian products include the amalgamation of traditional and contemporary materials and knowledge, and the adaptation of traditional art to fit into a modern milieu which makes it marketable. Examples of the latter include the substitution of acrylic paints for the ochres formerly used in both desert and Top End paintings in Australia, and modern dyes for the lichen colours once used in the Canadian Arctic. In both countries these changes have caused much discussion and some concern. Failure to respect traditional knowledge and responsibility, as Scott-Mundine (1990) and Golvan (1990) amongst others have commented, has led to an increasing number of claims by Australian Aboriginal artists against companies who have infringed their copyright by using their designs in the mass production industry, notably on T-shirts. Failure to recognise that Aboriginal culture, like any other, is dynamic rather than static has led to unjust criticism that today's products are not authentic. Myers and others (1988), commenting in relation to Inuit print-making, point out that aboriginal people today are not working in a cultural vacuum, cut off from all the processes of socio-economic and cultural transformation which affect their daily lives. Like all of us they are continually bombarded with changes,

213

and accordingly they incorporate these into their art.

Art and craft producers in both Australia and Canada come from all sectors of the aboriginal community – old and young, male and female – and their work often involves a high level of co-operation. The industry has been a great educator and source of cultural regeneration. However in drawing on such a wide range of people it is hardly surprising that problems of quality control have been common. This problem has largely been dealt with by channelling arts and crafts products through the marketing organisations, the 'middlemen' lying between the producers and their customers.

In Australia and Canada aboriginal arts and crafts can reach their market in three main ways. Products can be sold directly by individual producers to purchasers; or by producers to aboriginal co-operative organisations which also provide them with technical support and advice; or by producers to outside dealers – museum and gallery owners and independent retailers, most of whom are non-aboriginal. Each of these methods gives producers different types of control over the payments they receive and also over the way in which their work is presented to the public. The first method, direct sale, would appear to offer the highest level of control. However many aboriginal artists may have very limited knowledge of the realistic price for their goods, or of market demand and it is all too easy for outsiders to take advantage of them. An example of this commonly seen in Alice Springs is the sight of an individual Aboriginal artist selling a painting to pay for a taxi fare or a bottle of grog.

In some aboriginal communities artists have banded together to form co-operative organisations through which they can employ someone to deal with the market side of the business, and who may also help in obtaining art supplies at reasonable prices. In Canada's Northwest Territories many of the co-operatives which later came together as members of Arctic Cooperatives Ltd (ACL) began as art and craft producer groups. ACL's 'Northern Images' stores, located not only in the north in Yellowknife, Whitehorse and Inuvik but also in Edmonton and Winnipeg, act as front windows for the southern Canadian market. They are fed in turn by smaller outlets run by the co-operatives themselves. The Cambridge Bay arts and crafts store, described at the beginning of this chapter, is one such outlet. It is there that goods are purchased from producers, and all the difficult decisions about whether to buy the product, what it is worth and whether to pay in advance or withhold most of the cash until after it is sold – are made. Co-op managers often find this part of their job particularly difficult because many of them are not expert at judging artistic merit. And, as the description of Cambridge Bay showed, there are often severe problems in shifting the goods. As a consequence co-ops, both with the local community and the umbrella organisations such as ACL find that too much of their capital is tied up in arts and crafts stock, and their immediate liquidity is adversely affected.

Government agencies, in both Australia and Canada, have played prominent parts in the arts and crafts industry, particularly in marketing. This has undoubtedly provided essential support which has stimulated production in very remote communities where otherwise it would have been extremely difficult for contact with the market to be established (Williams, 1985). However, while aboriginal people have on the whole gained through their relationship with government organisations, some problems can occur. In particular, these linkages can put too much power into government hands, leading to a somewhat paternalistic control over artists, not only determining what their earnings are but also what they produce. The experiences of the Inuit print-making industry demonstrate some of these pitfalls. In this industry separate Inuit communities manufacture limited editions of prints from original paintings for show in periodic exhibitions, often in southern Canada (Myers, 1988). The print shops, located throughout Nunavut, were originally supported through DIAND funding and many provided the basis for the subsequent growth of local producer co-operatives. Government monitoring of these operations was established through a watchdog committee, the Canadian Eskimo Arts Council (CEAC), which assisted with marketing but also checked on the quality of the products. Unfortunately Inuit people lacked confidence in the judgement imposed on them by the members of this committee, largely drawn from southern Canadians knowledgeable in the art world. Matters came to a head in 1987 when CEAC informed one of the co-operatives that a significant number of their prints, already in production, were to be discarded. The estimated loss to this co-operative was $5,000 per print. The co-operative responded by pulling out of the government support system and trying to go it alone. This proved to be almost impossible, and the result was the collapse of the local industry.

Further problems resulting from dependence on government funding concern the bureaucratic perception of arts and crafts organisations solely as commercial ventures, while the producers and others in the community may perceive the cultural value of the industry to be equally if not more important. Thus Canadian Inuit co-operatives try to keep their arts and crafts components going even if they make a loss, thus undermining the commercial viability of the whole co-operative (Isaacs, 1988).

Many Australian Aboriginal communities have also established organisations which support their arts and crafts industries. Examples include Buku Larrngay (Yirrkala, Arnhem Land), Maruku (Uluru, Central Australia), Waringarri (Kununurra, Kimberley) and Warlukurlangu (Yuendumu, central Australia). These vary markedly in their mode of operation, their policies and the products with which they deal. Maruku, for example, is located within the ranger station complex at Uluru and thus, unlike most community arts organisations, has constant contact with a large number of tourists. Many are very short-term visitors – seeing the Rock in a one-day round trip from Alice Springs – and they are interested in buying small portable souvenirs rather than fine art. Although Maruku sells some expensive items and deals efficiently with credit card purchasing and packing for overseas mailing, its prime sales are in smaller craft items, such as carved wooden figures. Maruku's other unusual feature is that its staff travel to over fifteen Pitjantjatjara communities scattered throughout a huge region within three states, Western Australia, Northern Territory and South Australia. The logistics of its purchasing and payment systems are, not surprisingly, quite complex and decision-making is difficult. It would, for example, be virtually impossible to collect all the Maruku producers together in one place to discuss company policy.

Warlukurlangu, in contrast, represents only the artists of Yuendumu and its outstations, and is dedicated primarily to the production of paintings for the fine art market. Although Yuendumu people, like other Warlpiri, maintained their traditional artistic skills and knowledge within the ceremonial framework – through ground, body and artefact painting – it is only within the last few years that they have transferred these skills on to canvas with acrylic paint. However artists in this group have rapidly built up a reputation for spectacular and innovative art, using the western desert 'dot' tradition but often unconventional in both colour and the representation of key features. In 1986, for example, a Yuendumu painter won the annual Northern Territory art award with a painting which incorporated the 'dreamtime' story of the Pleiades, seven *napaljarris* (sisters) from his own country on the edge of the Tanami desert, with the Northern Territory flag and desert rose symbol – a combination which he presented as that of 'black and white' unity. This and many other Yuendumu paintings have found their way not only to Australian galleries but are also exhibited in the United States and in Europe. Warlukurlangu as a result is concerned not so much with dealing directly with customers but more with marketing through its linkages with galleries and museums, mostly interstate and overseas rather than local. The organisation also, as Lennard (1990) describes, plays a key role in providing artists with raw materials such as canvas and paints, and advises and supports both those who have already built up their reputation as artists and those who are making their first tentative steps in the business.

Artists and craftspeople also sell to dealers who travel independently around remote communities. Before the establishment of community organisations such as Maruku this would have been a prime way of linking producers to the market. At times this type of marketing was prone to exploitation, although most of the dealers, often from galleries in towns, offered fair prices and had often built up a very good rapport with the artists within their 'beat'.

Whatever the marketing strategy the question of pricing and return to the artists is of prime importance in both fine and tourist art. Outsiders often assume that all dealers are 'rip-off merchants' and that artists should be getting at least 80 per cent of the final selling price. Unfortunately this level of return is rare. Not only are many of the producers living and working a

Plate 6.7 Anmatyerre ground painting in ochres on sand, Mt Allan. These paintings, used as centrepieces for ceremonies, are the prototypes for contemporary commercial fine art from the central Australian desert

Plate 6.8 Warlpiri painting, produced for Warlukurlangu Artists in the Yuendumu camp

long way from their markets, but their paintings and carvings, like the soapstone musk ox at Cambridge Bay, may not be instantly saleable. Yet they want to be paid as soon as they dispose of the product. Wholesalers and retailers such as ACL, on the other hand, cannot afford to pay artists in full until after the product has shifted. Often a compromise has to be reached. This might mean giving the artists an up-front payment which is conservatively rated. Ultimately the artists may get less than they expected. Factors such as these underlie Altman's (1990) figures which show that Australian Aboriginal producers in 1987–8 received about 38 per cent of the total sale value of the industry, estimated at around $18.5 million. This would obviously vary according to the type of product. His estimates for a group of eleven major art centres, several of which operated in fine art, gave a return to producers of over 70 per cent (Altman, 1990: 9).

Infrastructure improvements from tourism

Tourist development means better roads, improved communications and the growth of better services in all sorts of ways. These things are often self-perpetuating – once the roads improve the visitor numbers increase and yet more facilities are needed. Aboriginal people have undoubtedly benefited from such changes. It has helped to break down isolation, and has also made it easier for them to make use of facilities available elsewhere. The development of the commercial sectors in towns like Yellowknife or Alice Springs, for example, is partly a product of the growth of tourism; it has also been appreciated by aboriginal people, not only those living within those towns but also those from the surrounding regions. Sometimes there have been specific gains because of the tourist industry. *Anangu* in the Mutitjulu community beside Uluru, for example, not only gained title to the 'Rock' and a strong voice in park policy, the agreement also gave them significant boosts in federal government funding for new houses and essential services.

Tourism and the subsistence economy

Although on the whole the tourist industry and its ancillaries seems to provide commercial benefit to aboriginal people in remote communities there is one important area where they may be economically disadvantaged – the subsistence economy. Disruption of the subsistence economy can occur in a number of different ways – destruction or degradation of natural habitats by uncontrolled tourist activity, with a resultant decline in the value of natural resources for hunting, gathering, fishing and trapping; and involving aboriginal peoples in tourism at levels which are so great that they prevent them from playing their full parts in the traditional subsistence economy. With recreational fishing in both Australia and Canada and with the big-game hunting industry in Canada tourists compete with aboriginal people for

218

the same resources. However while the tourists are interested in these resources for recreational purposes, aboriginal people are primarily using them for subsistence purposes. Unless the aboriginal peoples themselves exercise meaningful control on resource use, it is often the tourist demands which will take precedence. Once again, aboriginal ownership of the land and the resource is crucially important.

Evidence that employment in the tourist industry destroys people's ability to carry out subsistence is inconclusive. While people who take on wage jobs, of any type, obviously have to balance the monetary advantage against the loss of freedom to do other things, many of the activities in the tourist industry are quite compatible with subsistence. Part-time work is common, and arts and crafts production is normally carried on under self-employment. It is probable that most people feel that, so long as they are careful in selecting their job, they can successfully balance work with tourists with other activities.

SOCIAL AND CULTURAL ISSUES

A wealth of social and cultural issues must also be considered in examining tourism and aboriginal development. In contrast to the economic issues many of these seem to have negative effects – compromising cultural values, becoming a tourist spectacle and losing privacy; the growth of consumerism, antisocial behaviour such as alcoholism and divisions within the community which inevitably result from increased cash incomes and access to disruptive elements of non-aboriginal lifestyles; loss of land through tourist developments, and resultant intrusion of tourists into aboriginal communities. There can be some positive effects – regeneration of a wide variety of traditional activities; and spin-offs from educating tourists about aboriginal life and culture, in ways which enhance mutual understanding in a much broader sense.

Control of tourist access to places of spiritual significance is usually the first request of aboriginal traditional owners when they discover that parks and recreational ventures are planned for their country. Where they own the area, as with Uluru, Kakadu and Nitmiluk, they can exert control. Elsewhere their requests may be ignored and even if negotiation with park development authorities occurs the final solution may be a compromise which is barely acceptable. Pressures to allow tourists into sites can be very strong. Studies which I have conducted in the Kimberley region of Western Australia, for example, reveal that the non-aboriginal tourist operators are pushing very hard for the opening up of caves with rock paintings in areas accessible to the road. On the surface this request seems a reasonable response to continual comments made by visitors to the region that they want to see Aboriginal paintings. Nevertheless it does present problems, not least because according to Aboriginal cultural beliefs many sites can only be visited by either men or women but not both. As Aboriginal men say, they should not be allowing non-Aboriginal women to see paintings which would be forbidden knowledge to their own mothers, sisters and daughters.

Aboriginal artists and craftspeople may have to decide how far they can go in presenting some of their own cultural beliefs for tourist consumption. In producing paintings and carvings artists are also making very clear statements about their culture, their souls, their identities and their feelings about what has happened to them throughout their whole history of contact with the non-aboriginal world. These are things which might not otherwise have come into public knowledge and if an artist errs in presenting such information he or she may find themselves denigrated by other aboriginal people in their community. The social pressures arising through this type of experience can be quite severe.

Instances of unwelcome intrusion into aboriginal lives as a result of tourist development abound in Australia. These have been most profound where tourist numbers have been large, and where there has been notable lack of sensitivity on these issues. Photography, including commercial filming and taking individual shots, has often formed the core of such problems. Stories of tour buses travelling through outback Australia with the windows wide open so that visitors can easily take pictures without asking permission of their subjects are common. The Mutitjulu community at Uluru was particularly vulnerable to such treatment before the current park management system was in place. At that time the community had the only retail outlet and petrol bowser in the park, tourists camped next to the settlement and there was a constant flow of visitors through the small town. Particular incidents, such as a commercial TV documentary which portrayed Mutitjulu people as drunken and dirty but which was made without permission at dawn when people were just awakening from sleep, have had long-lasting effects. Aboriginal outrage over this type of treatment undoubtedly explains the strong line which the Pitjantjatjara Council, and Mutitjulu in particular, now take over requests to film at Uluru. The fact that the Mutitjulu people can now exercise control through declaring their community off-bounds to casual tourists has clearly been of great importance. Undoubtedly Canadian aboriginal groups will also find themselves facing these kinds of pressures as their tourist industry grows to larger proportions, and the experience at Uluru is one which provides a timely warning.

Increased cash incomes from tourism, as from mining, have had some negative social effects in aboriginal communities. These have included greater consumerism, particularly of goods which had not before been readily available but which tourists demanded, and an increased potential for social divisiveness caused by unequal access to the tourist dollar. However the effects here may be less significant than those of mining royalties because the financial gains to tourist-affected communities are, by comparison, quite small. The traditional owners of Uluru, for example, share an annual rental fee for the park of $75,000, and 20 per cent of the gate monies. In 1991/2 this would have amounted to about $575,000 in all, or about $3,500 per head (Australian National Parks and Wildlife Service, 1991). Compared to the comparatively large sums of money which have gone to the Gagadju people from the Ranger uranium agreement, this does not offer a substantial nest-egg for investment or community improvement. In the case of Auyuittuiq, where the park's establishment is part of the Nunavut settlement but where the land itself is still held by Parks Canada, no rent is paid. However it could be argued that payment is still occurring – in the form of a portion of the payment of $580 million which is part of this claims settlement. This obviously offers potential for investment.

Some aboriginal people have lost land through tourist development, both for resorts and for parks. In past times this occurred with little or no consultation between the developers and aboriginal groups and it may now be very difficult to compensate for these actions. However more recent decisions over such projects are, as discussed earlier, much more likely to involve discussion of all issues and possible alternative solutions. As a result, although loss of land is still occurring, there are possibilities for participation in the project, or for agreements which allow for controlled subsistence harvesting within park boundaries. Many have come to feel that the advantages of having land officially protected once it is declared a national park may outweigh the disadvantage from losing it.

Aboriginal people are increasingly coming to see that they can overturn some of these damaging social effects by using their contacts with tourists to educate non-aboriginal people about aboriginal matters. Initiatives of this type include controlling the products which are sold as souvenirs, as described for the *anangu* kiosk at Uluru ranger station; and controlling the information that tourists are given. Once again, *anangu* have been able to use their ownership of the land to exert a strong influence. Aboriginal people living near Uluru have for many years been conscious that tourists were often fed on a rich diet of misinformation when they were travelling within the park. Incorrect stories about places around Uluru, about its significance for aboriginal people and about aboriginal people in general were related by bus drivers and tour couriers. Some of the comments made were blatantly racist. *Anangu* have now responded by developing a detailed series of notes on all aspects of the park, its cultural value and its environmental character-istics, and distributing them to all tour companies bringing people to Uluru. They have also discussed organising training courses for couriers, and possibly some type of surveillance to ensure that couriers do their jobs properly. Occasionally, when they have had evidence of particularly poor performance on the part of guides, they have banned certain individuals from entering the park. Measures such as these cannot ensure that the misinfor-mation problem is eradicated. That would probably only occur if *anangu* themselves wanted to be the couriers. However the scheme has been received sympathetically by many of the companies concerned, and potentially should help to improve non-aboriginal understanding of aboriginal matters.

TOURISM AND ABORIGINAL DEVELOPMENT

The role of the tourist industry in remote area aboriginal development varies according to the approach taken. In general, large-scale tourism, involving substantial numbers of people, highly organised systems of transport, the establishment of big resorts and the provision of a wide range of recreational and entertainment facilities does not provide an appropriate scenario for participation by aboriginal people. Not only do they lack the necessary skills and experience but many would find the situation intimidating and very intrusive. Small-scale tourism, particularly tourism which concentrates on environmental and cultural aspects about which aboriginal people have particular knowledge, can offer attractive opportunities. Because of the small scale the possibilities of controlling these operations to accord with aboriginal perceptions of appropriate tourist behaviour are much greater.

Of all the enterprises associated with tourism, the arts and crafts industry seems to offer the best opportunities for aboriginal people. This reflects both its reliance on their own skills and knowledge, and the fact that the industry can be operated with highly flexible production systems. This allows people to work as artists and, at the same time, continue with other activities.

7

ABORIGINAL COMMUNITY STORES AND DEVELOPMENT

It could be argued that, while the resources of remote parts of Canada and Australia – the gold, uranium, oil and gas; the cattle and furs; the attractions of these regions for tourists – are all essential to development, it is ultimately the people themselves who form the most important resource. This means that the maintenance and improvement of the living standards of that population is itself vital. In other words, service provision becomes an extremely important part of the whole equation. Services such as education, health and housing, along with communications and other components of the infrastructure are fundamental. These not only give local people the necessary support to improve their skills and to take their part in technological change and advancement; they also provide them with jobs and enhance their incomes. They are thus an integral part of the remote area economy. One could argue that, since the people belonging to these areas have shown that they are strongly determined to stay there even when economic difficulties arise, the contribution of services to the economy is vitally important. It is certainly the most constant.

However most services are not in themselves expected to contribute to the process of economic development. Even with today's policies of 'user pays' they are not designed principally to generate profits. There is however one prime exception: retailing, an activity which provides a vital service in every community where people are part of the cash economy; but also an activity which, unlike health or education, is normally privately run and where profit-making becomes a prime aim. All but the smallest aboriginal communities in remote Australia and Canada have retail stores, where people can buy essentials such as food, clothing, tools, furnishings and whatever exotic items they want and can afford. They not only provide an essential service for the community but they are also expected to generate financial profits and provide a living for the owners. This gives them a dual purpose. In the case of aboriginal communities that dual purpose has been a source of conflict, because of the pressure placed on the store to demonstrate that economic enterprise is an important avenue to successful modernisation and progress in the ways of the industrialised world. The retail industry therefore provides a further important example of the problems encountered in promoting aboriginal development.

THE STORE IN ABORIGINAL COMMUNITIES

Friday shopping (Young, Fieldnotes, 1978 and 1982)

'Smoko' time, the morning break for all workers in a central desert Aboriginal town, is when everyone – young or old, men and women, with or without cash to spend – flocks to the shop. This happens every day of the week. But on Friday pay days the activity inside and outside the community store is particularly hectic.

The shop is seething with people. It is an 'outback' variety store, selling basic ranges of fresh, packaged and canned foods and an extended range of clothing, tools, camping gear and household equipment. Although basically a super-market, it is dominated by its large freezers, essential for the storage of all fresh items in this climate. Consumerism, similar in form to that apparent in any small Australian town, appears rife. People are filling trolleys with frozen, fresh and packaged foods and, depending on their finances, clothing, tools and household equipment, electrical goods, blankets and even furniture. What they buy reflects both basic needs, such as bags and drums of flour, packets of tea and sugar, meat; and also what they see others buying. The flow of conversation about what to try is continual, particularly if the fortnightly road train has recently delivered its cargo of fresh and frozen foods from Alice Springs. Then the choice is wide: cakes, cheese and fruit and vegetables, as well as fresh bread and milk. At other times it will be much more limited and if the road train is overdue the shelves will be half empty and the freezers will probably contain only a few packets of less favoured items such as frozen legs of lamb.

The shopping process is slow. Huge queues clog up checkouts, partly because of the quantity of goods being bought but also because customers, and some of the checkout operators, need time to work out their cash transactions. Some customers find that they have overspent, and have to negotiate with the checkout operator over which goods to leave behind. Others bargain for credit, using their pay to cover last week's purchases and beginning the fortnightly process of using the store as a 'bank' all over again. Occasionally a checkout operator signals to the store manager that she wants to be relieved from her position. She has spied members of her own family, wheeling laden trolleys and making straight for her, and is well aware that she may be put under very heavy pressure to turn a blind eye if they do not have sufficient money to pay for their purchases. All of these pressures are a strain on store staff and sometimes tempers become frayed.

Outside the shop other forms of frenetic activity take place. These are, to a large extent, quite different to what one would find in the vicinity of a town supermarket. The small mall is crowded with people, many sitting on the ground in groups which roughly correspond to extended families. Shoppers join these groups as they emerge and immediately begin sharing out some ready-to-eat purchases – fruit, cakes, soft drinks and pies from the take-away. Children in particular clamour for these, but everyone else expects a share, even if they themselves have nothing to contribute because they have no cash to spend. Other goods, packets of tea or tins of meat and even newly purchased

Plate 7.1 The old shop at Willowra

clothing or tools, are also shared around. Greeting friends, talking and gossiping all occupy the time and, while workers have to leave when smoko ends, others may well stay outside the shop for an hour or two. Arguments and fights, often related to sharing of goods or cash, are not uncommon. Eventually the shopping process, including its conventional commercial elements and its less conventional social interaction aspects, is complete and people leave for their houses and camps, hitching lifts whenever possible. The exhausted store staff, Aboriginal and non-Aboriginal alike, heave sighs of relief, do their own shopping, and close down for the weekend. The takings on a particularly heavy Friday's trading might well amount to 40 per cent of the store's fortnightly earnings.

The coming of the barge (Young, Fieldnotes, 1979, 1982, 1984 and 1989)

Extremes of climate, such as the monsoonal wet season in Australia's north or the winter in Arctic Canada, inhibit land transport to many remote communities and force stores to obtain their goods in other ways. For remote communities situated on the coast or on navigable rivers, such as those on Australia's Gulf of Carpentaria or Canada's Mackenzie River or Arctic shores,

225

the coming of the barge is an important community event. It means the arrival of all sorts of heavy goods too expensive to come by other forms of transport – building and construction materials, vehicles and fuel supplies. Above all it means the replenishment of the stock of the community store.

The arrival of the barge means at least a day of excitement and hard work, not only for store workers but often for others who willingly help in an unloading operation which must be completed as quickly as possible. Every useful vehicle – trucks, tractors and trailers and, in Canada, the ubiquitous four-wheel scooters which are the summer equivalent of the snocat – is pressed into action to carry boxes from the barge landing to the shop. Other people help to store and stack in the store's warehouse, and throughout all the proceedings managers check lading bills and monitor the condition in which their orders have arrived. This is vital. It is common for goods to be unloaded at previous ports of call, and also, particularly in north Australia where the barge is the vital link during the monsoonal season when cyclones are common, for goods to have suffered in rough weather. It is also common for orders to have literally 'missed the boat', a particularly serious situation in northern Canada where barges can only come during the brief summer.

The coming of the barge not only heralds the replenishment of depleted stocks in the store, or the arrival of long-anticipated hunting equipment. It also, particularly in Canada, marks the rhythm of the seasons. On the Mackenzie River people begin to discuss the barge's arrival as soon as the ice begins to break in June. The barge means that the short northern summer has arrived. In this part of the world, where water access is only possible for about three months in the year, only one or two barge visits are possible per season. Unloading the barge is also a very important social activity. As with shopping, it is an occasion for meeting friends and talking; and what the barge carries is soon common knowledge. People know what prized foods have arrived or whether the new consignment of TV sets or refrigerators is here. They start to plan their purchases even before these reach the store.

What would happen if the barge did not come, or if vital items were left behind? The only alternative for many places is air freight. The prohibitively high costs of this method of transport are really only acceptable for high-value goods, such as fresh foods or clothing; not for drums of flour or canned goods. Average prices in these shops can be up to 35 per cent higher than those in local regional centres like Alice Springs or Yellowknife and perhaps double those in Sydney or Montreal. In 1981 the Hudson's Bay Company, for example, stated that, in 1981 the freight on a five pound bag of potatoes sold in the isolated Inuit community of Pond Inlet amounted to $5.15, 73 per cent of the selling price of $7.10 (DIAND, 1984b). Without the barge the differences would be even greater. In communities with residents whose per capita cash incomes are only about half the national average these are very important considerations. Relying on the barge also, however, poses other difficulties for the stores. They need large spaces for bulk storage, spaces which, particularly in Canada, may need to be heated to maintain a particular temperature, and which always require careful maintenance to guard against the depredations of mice, rats and other vermin. Barges could only be replaced if air freight costs were slashed, as has occurred in the United States where special rates offered by US Mail have allowed air freight to compete in Alaska. But in Canada and Australia there are as yet few signs that this will happen. The barge, for those running the shops and those who buy from them, still remains a lifeline.

Plate 7.2 Supply barge on the Mackenzie River

Plate 7.3 Arnhem Land supply barge

227

Store competition (Young, Fieldnotes, 1984)

Which store to go to; the 'Co-op' or the 'Bay'? This is a question which continually confronts the aboriginal residents of isolated Canadian communities. It is not posed to their Australian counterparts who, unlike the Canadians, have never been part of a commercial empire like that of the Hudson's Bay Company. Australian communities not only lack trading posts but also have few non-aboriginal owned stores.

The choice between the Co-op and the Bay is hard. The Co-op belongs to the community, most if not all of its staff are aboriginal Canadians and it usually operates in a less formal way which people can relate to. It may also sell some prized local items such as country foods. But it is not necessarily the first choice of the customers. Why?

Firstly, the Co-op is often more expensive. The 'Bay', and its successor Northern Stores, has a huge organisation behind it. It relies on the savings made by Canada's principal retailing chain through bulk buying and bulk manufacture of its own custom-made products, and a transporting system which aims to be as cost-effective as possible. The Co-op, in contrast, relies on a support system provided by companies such as Arctic Co-ops Limited (ACL) or Federation des Co-operatives du Nouveau-Quebec (FCNQ). These companies are smaller, less wealthy and highly dependent on government subsidies for their very survival. Moreover ACL and FCNQ are not only concerned with supplying goods for isolated aboriginal Co-ops. They are also involved with the training of aboriginal store workers, in the hopes that one day all the Co-ops which they serve will be fully staffed and under the control of their aboriginal owners. These efforts cost a great deal, both in human resources and in financial terms. Inevitably customers bear some part of these extra costs in the prices which they pay in the shop.

Secondly, people have been using the Bay as a trading post for generations. Although, with the virtual collapse of the market for furs and pelts and the proliferation of local aboriginal-owned Hunters and Trappers Associations, this service is less important than before, old habits die hard. Some components of the trading-post tradition are retained at the 'Bay' – credit is still available, particularly to hunters bringing in pelts. Co-ops may not be able to afford to offer credit. Thus people may well continue to use the 'Bay' rather than their own store and competition, perhaps to the ultimate advantage of the customer if not to the survival of the local Co-op, is well established.

What of Australia? There the community store often exercises a complete monopoly, free to impose whatever stocking policy or price system it wishes, and often safeguarded by the fact that, with Aboriginal ownership of the land, no competitors from the outside are likely to be given permission to establish themselves. They are like the retail arms of the Co-ops without the 'Bay'. Some work entirely on their own and some are supported by wholesaling and management training organisations, smaller versions of ACL and FCNQ such as the Arnhem Land Progress Association (ALPA), Yanangu or Anangu Winkuku Stores (AWS). Exercise of the right kind of control by the community, under these circumstances, becomes vital.

Plate 7.4 Hudson Bay store, Cambridge Bay

RETAIL STORES AS AVENUES FOR DEVELOPMENT

These descriptions highlight many features common to retail stores in remote aboriginal communities. They underline the problems caused by physical isolation from main sources of supply for goods. Isolation from external markets is crucial, because it inhibits the enterprise from expanding to meet demands greater than those of the local resident population. The harsh environments and difficult terrain in which many of these communities are situated also present problems for these retail stores. Their operations are affected by some features of the economic and social structures and functions of the communities themselves. These include gross variations in trading patterns; limited capitalisation and investment; and poor managerial and financial experience which, coupled with customers whose buying power is restricted by their low incomes, hinders the potential of these enterprises to generate significant monetary profits. Stores also contribute to community well-being because they provide jobs and give people a varied experience which is useful in other sectors of the workforce; but this role is expanded to a level where it may undermine the store's potential to make money. Aboriginal social organisation and behavioural norms also have an effect on the operation of these enterprises. These characteristics together raise an important question. Are such stores there to provide a service to the community? Or are they there to earn money for the community, as a

229

demonstration of its increasing effectiveness in modernisation in ways which will tie it more closely to the industrialised society of which it is a part? This question, which I have raised on a number of occasions in the past (see, for example, Young, 1982; 1984; 1987b; Ellanna et al., 1988) can only be understood through examining the contemporary economic and social characteristics of stores against their historical background.

Aboriginal community stores as enterprises: the historical context

In both Canada and Australia stores have been focal points of contact between the aboriginal peoples and the non-aboriginal incomers – traders, government officials, missionaries and others. They have been the means of introducing people to goods and chattels such as rations of tea and sugar, blankets, rifles and modern clothing and as such have been a major instrument of assimilation. As trading posts they have introduced people to the concept of exchanging labour and goods for cash and, particularly in Canada, they have benefited non-aboriginal people by allowing the collection of valuable items for which external markets were known to exist, notably furs and pelts. In this sense they epitomise the exploitation of aboriginal society by representatives of the non-aboriginal incoming groups. In other ways they have helped aboriginal people, providing them with a wealth of goods which have made their lives easier and have increased their living standards. Stores have also, more recently, been perceived in commercial terms as a way forward, a channel through which aboriginal people could begin to make the progress which conventional attitudes towards development expected of them.

Australia

Stores in remote Aboriginal communities in the 1950s and 1960s, run by mission and government authorities, had a number of principal aims, most of which accorded with assimilationist policies. These included introducing Aboriginal people to western foods and discouraging them from subsistence activities; training Aborigines in western ways including what foods to choose, how to prepare food and the management of cash. The 'proper' behaviour for the ritual of shopping, including queuing conventions and tidy dressing were, as Middleton and Francis (1976) describe for Yuendumu in the 1960s, seen to be important parts of this training. Stores also gave Aborigines opportunities to join the workforce and adopt non-Aboriginal work ethics. Many contemporary stores have been based on such establishments and therefore it is scarcely surprising that many of these elements are still present. Most importantly these assimilationist aims are firmly entrenched in the minds of many non-Aboriginal people concerned with the store business, from managers to government and mission administrators.

Since the 1970s Aboriginal community stores have been given an additional role – that of being financially successful enterprises. This role is an outcome of the adoption of self-determination policies – that Aborigines should be controlling their own services and enterprises, and should be striving for greater economic independence from government and other authorities. Economic enterprises were promoted as a prime channel for achieving these goals. In most remote communities, the store was the only enterprise already in existence. These were quickly transferred from existing government or mission control to Aboriginal ownership as a direct demonstration of this new development policy. Most came under group rather than individual ownership, being placed under the auspices of community councils, social clubs or groups representing all community members. Some were set up under co-operative structures but their operations, in terms of earning and paying out dividends to members have rarely conformed to this. However collective decisions on how the stores should be operated, made by elected boards of directors, have theoretically provided the appropriate framework for democratic decision-making. These structures essentially reflect external, non-Aboriginal concepts of the role of community stores in development.

As I have described in detail elsewhere (Young, 1984), when measured according to criteria such as financial profitability, provision of employment and of opportunities for gaining experience in management, the actual performance of these stores has often been disappointing. Reasons for this, described below (see p. 234), demonstrate that policy-makers and bureaucrats failed to understand and respond to the needs of the Aboriginal community. They rarely consulted effectively to determine how Aborigines, as opposed to others, perceived the role of the store; they failed to recognise that the store was both a community service and an economic enterprise; and they failed to examine the institutional history of these stores, and to recognise the extent to which this continues to affect how they operate today.

Canada

Contemporary aboriginal community stores in remote parts of Canada share many common features with their Australian counterparts. However significant differences, stemming from forces operating during their establishment phase, also exist. Most importantly these include the ubiquitous presence, over many decades, of the Hudson's Bay Company; and the integration of these community stores with co-operatives, involved not only with retailing but also with the marketing of local products such as arts and crafts, fish and other items.

Hudson's Bay Company trading posts, set up since the establishment of the company in the seventeenth century, were initially concerned solely with co-ordinating the activities of independent trappers who provided them with the furs and pelts on which their business was based. In more modern times their trading role has been combined with roles very similar to those discussed earlier for Australian Aboriginal stores of the assimilationist era. As Ray (1990) and others have described, the Hudson's Bay Company effectively tied aboriginal people into a trading system within which the profits went mainly to the company rather than to the trappers. This made aboriginal trappers very dependent on the cash economy and undermined the independence which had been a hallmark of their richly endowed hunter-gatherer economy. In 1987 the Northern Stores Division of the Hudson's Bay Company had 178 retail outlets throughout Canada's north, in both aboriginal settlements and resource towns. Unlike some other components of the Bay's operations Northern Stores had remained profitable and in that same year it was sold to meet some of the debts of the company's struggling southern operations. The purchaser, North West Company, was a new company formed by the former management of the Hudson's Bay northern outlets. Thus, although this sale ended an era in the history of Canada's northern development (Bunner and Gallagher, 1987), the operation has continued under the control of people who were brought up in the Bay tradition.

Until the mid-twentieth century Hudson's Bay stores, and some small retail outlets run by mission groups, provided for the day-to-day needs of Inuit and Indians in the north. However perceptions that aboriginal people should be deliberately encouraged to take a stronger part in the cash economy and increase their self-reliance led the aboriginal affairs bureaucracy to feel that the time for a change had arrived. Increasing demand in southern Canada for local goods from the north, including arts and crafts and fish, provided the necessary stimulus. Locally based organisations were then established to co-ordinate these activities. These co-operatives, the first of which was set up at Kangirsualujjuaq (George River) in northern Quebec in 1959, were primarily producer organisations, co-ordinating the whole process of production – fish, making carvings from soapstone or clothing from furs and skins, or supporting artists through efficiently supplying them with raw materials – but they were also involved in marketing (Iglauer, 1966; Stager, 1982; FCNQ, 1988), mostly beyond the boundaries of northern Canada. Later, as money began to flow into the community, people wanted to buy more food and other goods. They also wanted to assert their independence. Most of the co-operatives then started retail stores. By 1990, for example, 36 out of 38 ACL member co-operatives in Inuit and Indian communities in Northwest Territories had retail stores. Many had operated in direct competition with the Hudson's Bay Company and are now competing with its successor, the North West Company.

As in Australia, non-aboriginal Canadians have had misconceptions about the roles and functions of aboriginal co-operative retail stores in their communities. Because of their locational disadvantage and the poverty of their customers they inevitably operate on a shoe-string and certain levels of government support are constantly needed to keep them afloat. This means that they have to grapple with government ideas about how they should be operated and developed, and brings sharply into focus the whole problem of combining two very different concepts of the whole process of modernisation.

Community retail stores: services or enterprises?

Almost everyone in Canada's high Arctic or Australia's central desert sees themselves as an expert on the problems of aboriginal community stores and on solutions to these problems. Questions on this topic draw the following replies:

> They're an absolute disaster, aren't they ? Full of out-of-date stock, rotten apples and things that nobody wants to buy! And they're usually going bankrupt because the manager's been dipping his hand in the till and the aboriginal people won't pay their bills.

These comments, in relation to the question of whether these stores are services or enterprises, are quite revealing. They expose two main perceptions, commonly held by most people outside the industry and closely linked to each other: firstly, stores are definitely enterprises; and secondly, their contribution to the community can only be measured by their level of profitability. Further discussion, almost certainly peppered with accounts of personal experiences of poor service and financial failure due to inefficient management and dishonest practices, lays the blame for the failure of the stores firmly on both aboriginal and non-aboriginal people within the community. Such ideas have some validity. However they only tell part of the story. They fail to acknowledge the broader multiple role of the store, with both economic and social functions, and they ignore the fact that its operations are affected by external as well as internal circumstances.

Economic considerations

The performance of community retail stores is, in straight commercial terms, affected firstly by the high cost of their operations; secondly, by problems of undercapitalisation and cash flow; and thirdly, by problems associated with staffing and management.

High operational costs are a function of both physical isolation and environmental disadvantage. They also occur because of other factors which make the commercial part of the business inefficient. Because of physical isolation freight costs add substantially to the price of goods. Small local markets prevent each individual store from being able to cut its costs by practising economies of scale. And environmental disadvantages, such as having to meet astronomically high fuel bills for heating in the Arctic or for freezer units and cool rooms in northern Australia also add substantially to costs. These costs are handed on to customers. However, because of the generally low income of consumers and the fact that, as owners of the stores, they exert an influence on its pricing policy, the level of cost which is allowed to flow through to the shopper may well be kept to the absolute minimum. This gives no margin for error. In unforeseen circumstances, such as failure in the transport system or an unusually prolonged wet season, the 'fat' in the system may be far too small and within a remarkably short time the business faces financial failure.

Problems associated with access to capital and cash compound these difficulties. When these stores came under aboriginal control they were usually housed in poor buildings, with very limited facilities for bulk storage or for keeping adequate supplies of fresh and frozen goods. Improvements depended on access to capital, theoretically available through their own profits or from loans negotiated through financial organisations. However profits are not only small, as discussed above, they are also extremely difficult to predict. These stores often have to pay for their stock before they sell it, and recouping the money means relying on customers who themselves have cash flow problems. As the brief description of Friday shopping shows, people trade very unevenly, when the pay comes in, and for the rest of the fortnight may well expect the store to provide them with credit. Almost inevitably this means that improvements in store infrastructure will be met from loans. Most of these come from government agencies because the store organisations are not attractive propositions for banks and private finance companies. This raises another issue, discussed below (see p. 235). Outside (government) agencies have come to exert unduly high levels of control over the operation of aboriginal community stores. Considering that there are likely to be some marked differences in the development agendas adopted by government and by aboriginal people, the owners of the stores, this suggests potential conflict.

Staffing and management problems further affect commercial operations. Aboriginal owners not only inherited or were forced to establish under-capitalised businesses but they themselves at that stage had little or no experience in management. The training which they received during the assimilation era limited them to the more menial tasks of unpacking boxes, cleaning shelves and fetching and carrying. Elevation to the job of checkout operator was rare, because it was considered that handling cash might put too much temptation in their way. Questions concerned with ordering stock, pricing goods and maintaining balance sheets were a mystery, certainly part of the white person's domain. Today much of this mystique remains. Although more and more aboriginal people have become involved in all levels of store operations, including management and finances, they still encounter problems in understanding these aspects of the operation.

This reflects the educational deficiencies of the staff and of the store directors, particularly older people, as well as problems in establishing and maintaining appropriate types of training and support. Cargo-cult type beliefs about the level of store profits and their subsequent use are still common. In Yuendumu in 1986, for example, the community was rife with a rumour that the store was taking $1.5 million every month. People were extremely upset because there was so little evidence that this bounty was flowing through to the community. Bitter accusations of malpractice, levelled at the management, the Alice Springs-based accountants and at the store's owners, the Yuendumu Social Club and its Aboriginal directors, followed. Eventually, following vocal exchanges at a community meeting, people were convinced that the figure was grossly exaggerated; it was in fact close to the annual store takings. One positive outcome of this experience was a promise from both the accountants and social club to systematically inform members of the community about the financial affairs of the store. However although problems such as this can be sorted out, they can have serious longer-term repercussions. They fuel suspicion and bitterness and foster feelings of illwill which are detrimental to community well-being as a whole.

Social considerations

Like other activities in remote aboriginal communities store functions and operations are strongly influenced by social factors. Some of these are part of the life of any small community, but others reflect aboriginal social structures and behavioural norms. As the description of Friday shopping showed, stores are very important gathering places and hubs for the exchange of information. In both Canadian and Australian aboriginal communities many people like to visit the store every day, even if they have little or no cash to spend. Such behaviour is no different from what one would find in any small town, or indeed in the suburbs of large cities. The implications for the store are that there should be space to accommodate these needs, and that the store keepers should accept that this behaviour will be part of the activity. In the Australian Aboriginal case that has certainly not always happened; people have been told to go away when they are not buying goods, and groups congregating outside have been asked to move on. With most stores now at least nominally under Aboriginal ownership, this is less likely to occur.

Forms of aboriginal social behaviour affecting the process of shopping and the operation of the enterprise include the sharing of food and other goods as part of the reciprocity systems which are so important in Aboriginal, Inuit and Indian societies; and the need to conform to traditional authority hierarchies, even when people are essentially working in a non-traditional situation. Traditional land ownership responsibilities may also influence who controls the operation of a shop and can lead to demands for rental payments.

Sharing of purchases from the store, described above, does not affect the operation of the shop. However when the same activities occur before payment the loss in store revenue can be considerable. 'Shrinkage', the loss of stock without payment, is a problem in many of these stores, and in fact is so widespread that many managers automatically build a shrinkage factor into their pricing system. It occurs partly through aboriginal store staff 'sharing' goods with customers to whom they are related. But it also reflects the operation of traditional authority. Most store staff are younger, educated people, and their elders exert pressure on them to give them goods without payment. This is why it is common practice, at least in Australia, for aboriginal checkout operators to ask for a break when they see members of their family coming into the shop. Such pressures are particularly severe on young women, who make up the bulk of checkout operators. Kinship and authority hierarchies also affect who is employed in the store, with the family members of store directors or other store workers often getting preference. This can have its benefits, as it usually results in harmonious staff relations. However it can lead to the employment of too many staff, and of staff who are less well qualified than other potential workers. Land ownership, of the actual area in which the store in situated, can also become an issue as people battle for control of what is seen as a valuable asset. Store directors may well be drawn disproportionately from families with traditional land holding rights over the store site, and other groups in the community who, because they have been dispossessed and pushed out of their country, lack these recognised traditional ties, are excluded. Thus, although the store is a co-operative or a community-owned venture, not everyone within the group necessarily has equal rights to control it or benefit from it.

Social considerations such as these are a necessary part of the operation of aboriginal community stores. They can and do have marked effects on the enterprise. These can be beneficial if they are used to make sure that the shop works in a manner appropriate to the type of community which it serves; but they are often detrimental to the successful commercial operation of the store. The dilemma is to provide a socially appropriate store (a service) that remains financially viable (an enterprise).

Solving the service/enterprise dilemma

Nowadays stores are absolutely essential to the lives of aboriginal people in remote communities. Failure brings universal misery and despondency, not only because of the inconvenience which occurs but also because it is an overt demonstration that aboriginal people have yet again failed to come to terms with the non-aboriginal world. They are usually made to shoulder most of the blame. Stores must therefore be kept going, and hence their commercial viability is vital. However they must also truly belong to the community, a culturally appropriate rather than an alien institution. So community control, including whatever forms of social behaviour are seen to be acceptable, must also be part of their make-up. These requirements pose difficult problems. Both Canada and Australia are still experimenting with solutions to these problems. The measures which they have been taking are similar. They include the establishment of centralised organisations designed to support stores in remote areas, not only through rationalising their supply systems but also in providing staff training and support for management. They also include recognition of the need for aboriginal control, not only of the community stores but also of the centralised organisations. Alternative approaches include the replacement of the community store or co-operative with a number of smaller businesses, individually owned and managed by aboriginal people.

Store support organisations

Canada

Canada's two main organisations which provide support for remote aboriginal community stores are Arctic Co-operatives Limited (ACL) and the Federation des Co-operatives du Nouveau-Quebec (FCNQ) (Figure 7.1).

Arctic Co-operatives Ltd

Arctic Co-operatives Limited (ACL) was established in 1982 following the amalgamation of Canadian Arctic Co-operatives Federation Limited (CACFL), a federated body charged with co-ordinating practical support for the general operations of NWT co-operatives, and Canadian Arctic Producers (CAP), a co-operative specifically concerned with the marketing of community aboriginal arts and crafts. ACL, with gross revenue in 1990 of over $42.2 million, is currently dominant in the Co-operative Movement in NWT, and a major business operation in the region.

ACL was established to meet service needs which had been recognised by aboriginal co-operatives already in operation. Effectively it is owned by its member co-operatives and is responsible to them for its operations. This gives the members a high level of control over ACL's operations and policy, and ties the service organisation in with the stores. It helps to ensure that ACL provides the kind of service which its members want. These ideals are framed in ACL's mission statement:

> to be the vehicle for service to, and co-operation amongst, the Northwest Territories co-operatives; hence, providing leadership and expertise to develop and safeguard the ownership participation of the northern people in the business and commerce of their country, to assure control over their own destiny.

(ACL, 1990: 15)

237

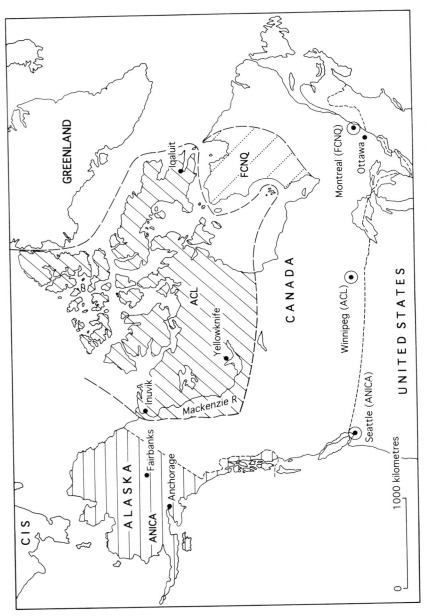

Figure 7.1 North America: regions served by ACL, FCNQ and ANICA

As this statement indicates, ACL's role goes beyond that of providing better services. The company also has an important political function, providing a strong foundation for supporting and fostering the growth of aboriginal identity in the geographically remote communities in NWT. All its current activities – helping member co-operatives with the purchasing and transport of supplies; providing them with management advice and with appropriate training opportunities; and, by bringing them together, helping them to deal more effectively with the formidable problems posed by remoteness and isolation – aim to fulfil that mission.

ACL's membership currently consists of thirty-eight co-operatives spread throughout NWT (Figure 7.2) and in addition it runs seven stores which retail arts and crafts, some of which are in southern Canadian provinces. It was originally located in Yellowknife, the territorial capital. However Yellowknife, in the extreme southwest corner of NWT, was not central to ACL's region, particularly the eastern Arctic area which focused on Iqaluit. It also, like all places in NWT, suffered from its isolation from the suppliers of all types of goods and services, most of which flowed from southern Canada. ACL's poor financial performance in the early 1980s focused attention on these issues and resulted in the removal of its operations to Winnipeg, where costs could be cut and where all its activities could be more easily rationalised. ACL's performance has subsequently vastly improved, with revenue increasing threefold in only eight years, from $14 million in 1982 to over $42 million in 1990.

In 1990 all except two of ACL's member co-operatives ran retail stores, and ACL had come to see the support of this service as its primary role. ACL head office processed all orders for store supplies and arranged for these to be transported to the co-operatives. Most orders went through the Canada-wide organisation Federated Co-operatives Ltd, of which ACL is a member. Transport was by air, river/sea, rail and road, depending on the location of communities and time of year. Water or land transport were preferred where possible because they were much cheaper; air, for anything except fresh or high-value non-bulky goods was to be avoided if possible. The short summer season, the only period when water or land transport is feasible, posed a major practical problem. Unless member co-ops bring in at least three-quarters of their supplies by water or land their retail prices will be too high for most of their consumers, most of whom have low incomes, to bear. They therefore have to pay in advance for a huge amount of stock, and can meet this bill only with ACL's assistance. ACL carries the costs for them by way of a large bank loan, guaranteed by federal government funding.

Although ACL is responsible for appointing its own staff it is not responsible for appointing co-op managers. That responsibility is carried by the board of directors of each co-op. In theory it would be quite possible for a co-op to appoint an inexperienced or dishonest manager and the results could be disastrous, not only for the co-op but also for ACL, a company

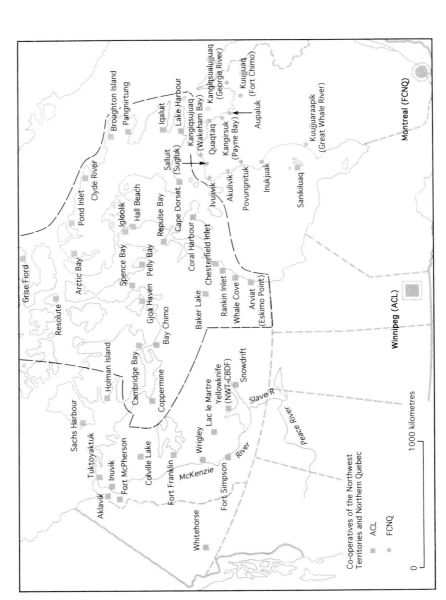

Figure 7.2 Canada: member co-operatives of ACL and FCNQ

which can only flourish if its members are operating efficiently. This has been widely recognised by the ACL's aboriginal directors who have proved quite capable of closing down co-ops which are not financially viable. As Stager (1982) has commented, this was seen as an unprecedented step in ACL's early days but subsequently people have become much more pragmatic about this type of decision. The need for good management, coupled with ACL's clear policy of ensuring that, as far as possible, local aboriginal managers should run the co-ops, has led to a strong emphasis on management advice and training.

ACL assists the retail stores by providing management support through its area manager program. Member co-operatives are grouped into regions each of which is served by an area manager who carries out regular visits. That person's responsibilities range from advice on all sorts of general matters to assisting with the recruitment of new managerial staff or monitoring the financial operations of the business. While the needs of each co-operative vary from time to time, ACL generally tries to allow as much local control as possible. At the same time its staff try to ensure that serious financial problems do not arise. Education and training (human resource development) therefore closely complements the area manager scheme. As Abele (1989) summarises, ACL received funding from the federal Special ARDA program to conduct a major training scheme between 1979 and 1983. This resulted in an additional twenty aboriginal managers or department managers within member communities, an achievement which exceeded the original forecast target. Unfortunately the momentum was not continued because of withdrawal of funding, not because the scheme was unsuccessful but because the funding bodies had other priorities. Nevertheless this effort undoubtedly helped not only to increase aboriginal involvement in ACL but also to raise levels of community awareness of what the organisation was doing. ACL has also tried to do this through extending its training schemes to others in the community who are concerned with running the co-operative, particularly the members of its board of directors.

Finally, ACL provides an accountancy service, used by almost 75 per cent of its members in 1990. Although members are free to choose whether to use these facilities, they are encouraged by their relatively low costs. If the co-operative is in financial difficulties, and hence imposing a threat to ACL's profitability, it may be forced to join the system.

During the last five years ACL has also become involved with the distribution of government loans to its member co-operatives. In 1986 the NWT Co-operative Business Development Fund (NWTCBDF) was set up with federal and NWT government capital of $10 million to administer loans and other financial services to member NWT co-operatives. The main initial federal contributors were the Department of Regional Industrial Expansion (DRIE) and DIAND, over 97 per cent of the contribution from that source. In 1990 NWTCBDF's membership included all but five of ACL's members.

NWTCBDF's operations have subsequently been financed through contributions from ACL and its members and by 1990 the fund has been able to lend over $42 million to its members. Borrowers have apparently become more and more reliable in repaying their loans; ACL (1990) has attributed this to a better understanding that the money in the fund essentially belongs to all the members – it is not a government hand-out.

General issues related to ACL's operations include the degree of aboriginal involvement and control in all components of the company; its role in fostering aboriginal unity and political development in the region; and the involvement of government, both federal and local, in its operations.

ACL, as an avowedly non-profit making co-operative responsible to its members, must disburse its surpluses in the form of dividends, shares or other forms, depending on decisions taken by its board of directors. Amounts vary greatly but in 1990 $500,000 was distributed in dividends and a further $250,000 was added to NWTCBDF. All ACL directors are aboriginal Canadians, each representing one of the seven districts into which the territory which it serves is divided. This theoretically and practically gives aboriginal people control over ACL policy. It may be more difficult to put that policy into practice. In 1984 most of ACL's managerial staff, especially at head office, were non-aboriginal. Members were concerned about the prevalence of 'white faces' in Yellowknife and also commented that lifestyles and working conditions in central office were superior to those experienced by staff of the member co-operatives. These comments were related to concern over how their payments for ACL's services, 7 per cent of their takings, were being used. Subsequently, with the extension of management training, more aboriginal people have been employed at higher levels in the company and with time these differences should be less noticeable. They do, however, indicate that the relationships between ACL and the member co-operatives have to be kept under scrutiny, so that conflicts are defused before they become destructive.

ACL's broader political role within NWT has grown through time but is still not clearly defined. One reason for this concerns its membership, which includes both Inuit and Dene/Metis communities and also town-based co-operatives. Inuit co-operatives made up over three-quarters of the 1990 members but Inuit interests are not the only ones which ACL represents. This potentially could be a source of conflict. However ACL has felt that it should assert its role as an Inuit representative organisation and has demonstrated this by successfully applying for representation in the recent negotiations for the Nunavut settlement.This was done in order to defuse any undue competition which may come from the regional development corporations which will be administering the $580 million pay-out from that claim settlement.

Finally, what of the involvement of government? On its establishment in 1982 ACL inherited the financial indebtedness of its predecessors, CACFL and CAP, both of which had survived only through the intervention of government funding. Between 1978 and 1983 the government established a Co-operative Development Fund of $15 million to bolster CACFL, CAP (and then ACL) and also FCNQ (see below). It was felt that the survival of the community co-operatives was absolutely essential for northern development. Following a review of this scheme (Stager, 1982) it was recommended that government assistance be rationalised to place emphasis primarily on transport and bulk-buying subsidies (for example the sea-lift) but some government input into ACL's operations still remains. Nevertheless the establishment of NWTCBDF has effectively decentralised the administration of government funding from DIAND to ACL, an organisation which is largely directed by aboriginal Canadians. This should improve the degree of aboriginal input into this important area of financing. It might also improve the efficiency with which funds are disbursed. In 1984 ACL management commented that severe delays in the receipt of government grants and loans could have very serious repercussions for their whole operation.

Federation des Co-operatives du Nouveau-Quebec

Montreal-based FCNQ, established in 1967 to support the newly developing co-operatives in Inuit communities on the northern shores of Quebec province (Figure 7.2), offers services very similar to those of ACL. And like ACL it is owned by its member co-operatives, only five at the time of its establishment and now twelve. This membership group, much smaller than that of ACL, is relatively homogeneous both culturally and in terms of geographical location. It is predominantly Inuit, and is wholly located within Quebec province. As a result factional interests which affect ACL's operations are less prevalent in FCNQ.

Although FCNQ's members all began as producer co-operatives they now run retail stores and can avail themselves of supply purchasing and shipment services, accounting and auditing assistance, and training and education in the management field. Services directly concerned with retailing – the co-ordination of orders, their transmission to suppliers and the organisation and transport of supplies from Montreal to the member communities; training for retail store staff; and auditing and financial services – account for a significant proportion of FCNQ's revenue, almost $4.5 million in 1987. Transport costs, for shipping during the summer but increasingly for the air-lift, are considerable and meeting the interest on the loans required to cover the purchase and movement of bulk goods to communities during the short summer period is a continuing source of anxiety. However, since the FCNQ co-ops are closer to their central organisation than are their ACL counter-parts FCNQ's transport costs are not so excessive. In 1982, for example, airmail postal rates to northern Quebec were only 13c/lb compared to 72c/lb in Northwest Territories.

243

Major issues concerning FCNQ are very similar to those affecting ACL – aboriginal involvement both in running the company and in working for it; the degree of government control over its operations; and its political role, particularly in relation to Native Development Corporations.

FCNQ always seems to have had a higher level of aboriginal employment than ACL, both in the co-operatives and in head office. Ever since the 1960s many FCNQ Co-ops have been managed by Inuit members of the community. Strong support from FCNQ, in the form of training, has been essential in maintaining this record. Training courses, in general store management, store accountancy, store employment and, more recently, hotel management, and paid for with federal government funding, are held every year. FCNQ's board of directors represents every member co-operative and all directors, except for FCNQ's general manager and members of the head office liaison staff, are aboriginal Canadians. In 1982 Stager commented that, while FCNQ directors had a good grasp of how the company functioned, local directors of individual community co-operatives were less familiar either with the co-operative system or with business practices.

Government assistance was important to FCNQ in the early years. In 1967/8 FCNQ derived 61 per cent of its income (then totalling only $114,000) from funds allocated by the Quebec government. However by 1971/2 the Quebec government share of the income of $397,000 had fallen to only 23 per cent and by 1987 this had dwindled to zero. Federal government funding, particularly during the five years from 1977 when FCNQ was a recipient, along with ACL, of funding from the Co-operative Development Fund, has also been significant and in 1978/9 22 per cent of FCNQ's income of $1.37 million came from that source. Stager's (1982) study recommended that federal government funding for FCNQ, except for the backing of the bank loan for purchasing and transporting supplies, cease as quickly as possible, and by 1987 government funding had virtually disappeared, accounting for only 0.6 per cent of the total income of almost $4.5 million. Over the whole period from 1967 to 1987 FCNQ has obtained 92.5 per cent of its income through its own resources, including funds from member co-ops, with the federal government allocating 5 per cent and the remainder coming from the government of Quebec province. This effectively gives FCNQ a higher level of independence than that achieved by ACL. Even FCNQ's successful bid for a loan of $3 million from the NEDP Cooperative Development Loan Fund in 1987 does not undermine this as this money, along with other government funds to back the transport bank loans, has to be paid back. Freedom from dependence on government obviously has important implications for meaningful Inuit control over decision-making in the company, a main priority for members.

The potential for competition between FCNQ, member co-ops and the Makivik Development Corporation has been an important political issue and careful negotiation has been needed to defuse possible problems. FCNQ came into existence before the completion of the initial negotiations between the Cree and Inuit and the Quebec government over hydro-electric development in the northern part of the province. Stager (1982) argues that the political experience which the Inuit gained while establishing their own co-ops and setting up FCNQ was extremely helpful in empowering them in subsequent discussions which led to the James Bay Agreement. However although the co-ops and FCNQ were obviously already making important contributions to social and economic development in the region they were not directly represented in the land claim negotiations. The James Bay Agreement subsequently led to the establishment of organisations such as Makivik, specifically charged with aboriginal economic development, but with no official linkage to FCNQ. Not surprisingly relationships between these groups have been somewhat strained. Stager suggests that aboriginal people identified much more closely with their co-ops and in the early stages saw Makivik as a rival institution. These tensions would presumably have been dissipated if the co-ops and FCNQ had taken part in the land claim negotiations, as ACL has been able to do with Nunavut.

A final characteristic held in common by ACL and FCNQ is the extension of their operations beyond that of retail service support, particularly in the area of arts and crafts marketing. This has affected their financial situations. As commented earlier, there is often a considerable time lag between the completion of a work of art and its sale to a buyer; meanwhile the producer wants to be paid. An additional problem, as far as ACL and FCNQ are concerned, is the desire of their members to pay for the retail services in kind rather than in cash – through trading them carvings to sell. In 1982 FCNQ's inventory included $3 million of carvings, an asset which substantially exceeded its total annual sales; ACL's assets in 1990 included an arts and crafts inventory worth approximately $2.8 million. Trading has not been easy. They have recently amalgamated that part of their businesses under a subsidiary company, Tuttavik, but that attempt to save costs has been thwarted by the downturn in demand for products. This is partly a result of recession, but has also been caused by the collapse of the market for goods made from skins and pelts, particularly sealskin. The successful banning of these and other products such as ivory has, as Wenzel (1991) graphically describes, been directly responsible for forcing many northern aboriginal artists to depend on welfare for survival.

Australia

Retail stores in remote Australian Aboriginal communities still have no centralised servicing organisations which are exact equivalents of ACL or FCNQ. This could be attributed to a number of factors (Young, 1993). Australia lacks a strong co-operative tradition such as exists throughout rural Canada; and, until relatively recently, Australian Aboriginal communities have only produced a few items, such as arts and crafts, which had to be traded to markets elsewhere. Producer co-operatives, as in Canada, were less necessary. Moreover many Australian Aboriginal community stores were not started by the Aboriginal community but rather by non-Aboriginal, largely government or mission, operators. Consequently they have not been part of a communal grassroots organisation, the characteristics of which are so graphically portrayed in Iglauer's (1966) story of the Inuit George River Co-operative.

Both government and mission authorities have strongly influenced the establishment and development of centralised service organisations catering for the needs of remote retail stores in Australia. It is only in the last decade that Aboriginal community groups have themselves set up their own support organisations, and even then this has required direct external, usually government funding, support. Current organisations include the Arnhem Land Progress Association (ALPA) (Darwin), Anangu Winkuku Stores (AWS) (Alice Springs) and Yanungu Stores (Alice Springs). All of these, compared to groups such as ALC, operate on a small scale.

Arnhem Land Progress Association

The Arnhem Land Progress Association (ALPA) currently owns and runs five stores in Arnhem Land Aboriginal communities: Galiwin'ku, Milingimbi, Ramingining, Minjilang and Gapuwiyak. It also provides 'consultancy' services, assisting the operation of stores owned by the communities concerned, in five other locations: Warruwi, Belyuen, Umbakumba, Warmun and Noonkanbah (Figure 7.3). With the exception of the last two, all of these communities are within the Top End of the Northern Territory.

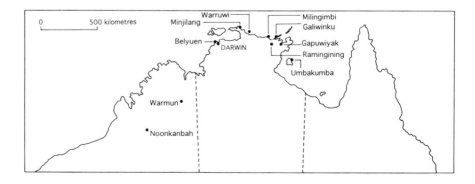

Figure 7.3 Stores served by Arnhem Land Progress Association

ALPA was established in 1972 by the Methodist Overseas Mission (the United Church of Northern Australia from 1972, and later the Uniting Church) as one way of fostering community development in accordance with the then newly adopted Commonwealth government policy of self-determination. Set up by the Mission as an Aboriginal organisation, it acquired ownership and responsibility for Mission Stores, retail outlets already operating in the Arnhem Land Methodist communities. From the beginning ALPA adopted a policy of operating its stores on strict principles of fair-dealing and honesty. While the original idea of ALPA, and the resultant structure of the Association emanated from the mission rather than from the Aboriginal people in the communities of Arnhem Land, its establishment did reflect Aboriginal opinions about how development might be fostered. The achievement of ALPA's goal of self-determination, in terms of Aboriginal control of ALPA's operations and Aboriginal involvement in all levels of employment, has presented some problems discussed in detail in Young, Crough and Christopherson (1993) and highlighted below (see p. 249).

ALPA offers the following services: bulk ordering stock, through G. and R. Wills; appointing store managers and providing support to ensure that they operate efficiently according to ALPA guidelines; encouraging the employment and advancement of Aboriginal staff, through training programs; providing an accountancy service, using standardised reporting procedures both within stores and between stores and the Darwin head office. It is independent of government funding, although the Aboriginal councils in the communities in which its stores are situated have periodically received government assistance for the construction of new store buildings and other capital development. Its surplus profits, derived from those earned by each individual store and from its investment earnings, are distributed according to a formula agreed upon by the board of directors. Recently in each financial year each community has received a sum equivalent to 20 per cent of their store profits plus 5 per cent rotating profit. The size of profit available for redistribution has expanded significantly during the last decade. In 1982 this amounted to $44,000 to six communities; in 1990–1, with only five ALPA stores remaining, the communities received over $187,000 altogether, the three with the most profitable stores, Galiwin'ku, Ramingining and Milingimbi, each receiving over $40,000. A total of $168,000 was distributed to the ALPA communities in 1991–2. Subsequent redistribution of these monies, both for community purposes and for individual payments, has been the responsibility of local organisations, generally the community councils. Some of the remaining, much larger, component of the surplus profit is used for development support, for example through schemes designed to assist other Aboriginal enterprises and to provide educational opportunities; and some has been invested in real estate.

ALPA's operations have, at different times, been extended beyond support services for retail stores. These have included a venture into the wholesale business and, recently, investment in real estate. Entering the wholesale business would seem to be a logical step for organisations which co-ordinate orders for large quantities of goods within a limited range. However ALPA's experiences in this area demonstrate many of the pitfalls which are likely to occur. ALPA's Darwin warehouse dealt mainly with variety goods and clothing, and could be used by buyers from non-ALPA as well as ALPA stores. During the mid-1980s ALPA tried to expand its activities and, along with establishing an East Kimberley office to meet training and service needs in the whole Kimberley region, made a bid for purchase of a local Darwin wholesale grocery company, Hickman's. Its bid rested on receiving Commonwealth government funding through the Aboriginal Development Commission and, because of delays in making the funding decision, was ultimately unsuccessful. Subsequently, in 1985, ALPA did gain control over another grocery wholesaler which, as Wells (1993) discusses, allowed them to offer their own and other stores a full range of appropriate stock. It was anticipated that this new ALPA wholesale company, North Centre West, would be able to attract the custom of a high proportion of Aboriginal community stores and that it would prove to be a very profitable business in the long term. However these plans failed to come to fruition. The small population of Australia's north, coupled with high freight costs and the fact that all major suppliers are in distant cities in southern and eastern Australia, make local wholesaling a risky business. In 1986 ALPA was forced to acknowledge the difficulties faced by their wholesaling operation and sold the whole business to Independent Grocers Co-op., a much larger southern-based operation.

ALPA has recently begun to make significant investments in real estate in Darwin. From the company's point of view this signifies a logical use of surplus profits, and it helps to ensure future commercial viability. It also, as the 1992 Annual Report stresses, provides the company with higher levels of investment return than would be obtained through banking. However many ALPA community members find the principle of investment decision difficult to understand. They are concerned that ALPA's profits, only part of which are returned every year to the communities, are being used for purposes, such as buying flats in Darwin, which confer little obvious immediate benefit to them. People have wondered why the money is not recycled into the stores instead, to improve buildings and equipment, or perhaps to allow for lower prices through increasing the range of freight subsidies. Such feelings suggest communication problems within the ALPA system, between the head office, the ALPA stores and the Aboriginal people concerned, both directors and community residents.

Since 1987 ALPA has sponsored a range of family-based enterprises, including a take-away outlet, a fuel depot, a video-hiring business, a local bus company and a fishing business. In all cases ALPA has granted loans for establishment, purchase of equipment and other initial costs. These businesses, as Wells (1993) records, have had mixed commercial success and in some cases little cash has been repaid to ALPA. Other difficulties have concerned competition for the funds, and the feeling that applications from some communities were being received more favourably than others. Galiwin'ku in particular, where the presence of the training school has ensured a strong level of management support, is seen to have been very favourably treated.

Decisions on all matters concerning the association are theoretically taken by ALPA's board of directors, all of whom except the chairman and executive director are Aboriginal. This appears to give Aboriginal people a high level of control over ALPA. In reality, technical matters such as those concerning the stock ordering systems, dealing with the transport services provided by Barge Express and by air freight, developing training curricula and the methods for delivering these, supervising the work of non-Aboriginal store managers and running the financial and accountancy services, are dealt with by the executive director and head office staff. Most of these are non-Aboriginal. Thus Aboriginal control in ALPA's day-to-day operations in Darwin is limited. This Aboriginal/non-Aboriginal split in responsibility for policy and technical matters is unlikely to change until Aboriginal people have been given the opportunity to acquire the necessary technical skills. It is not the only such split. The issue of whether to promote commercial success or low-cost retail services is also important, as discussed below (see p. 259).

Decisions regarding ALPA's policy, and particularly its function as an Aboriginal organisation, should very much be the responsibility of its Aboriginal directors and members. It is here that current issues of concern arise. As Crough and Christopherson (1993a) discuss at length, issues include the role of ALPA directors and their ability to communicate feelings and decisions both within the association and within their communities, the distribution of store profits both from ALPA and within the communities, and the issue of who exerts control over each community store and the employment of Aboriginal people within the association.

ALPA selects and employs its own store managers. They are chosen for their experience and efficiency and for their interest in and commitment to working with Aboriginal people in their own communities. Above all they are charged with providing as good a service as possible, and issues such as awareness of the need for fresh foods, and consulting closely with Aboriginal customers about their needs and priorities have always been stressed. The training of Aboriginal store employees, to give them experience not only in basic jobs such as checkout operation but also in costing, accounting and ultimately in management was a prime emphasis. ALPA established a storeworker training school in 1981, locating it in Galiwin'ku rather than in Darwin in the belief that the familiar setting of an Arnhem Land community with an ALPA store would be more appropriate than a city setting. Trainees combined their short courses at the school with on-the-job training, to which all ALPA managers were obliged to commit themselves. Courses were offered at all levels, from basic storeworker instruction to accounting and other modules which covered the various tasks which were part of the manager's job. The ultimate aim, as expressed in the ALPA Annual General Meeting report of 1984, was for Aboriginal managers to take over in all stores and for non-Aboriginal managers to be reclassified as support managers, assigned to understudy and advise their Aboriginal managers as required. These training schemes, at that time, were unique and were seen as the prototype for other bodies with similar needs. They drew their students not only from ALPA store staff but also from communities as far apart as the West Kimberley region of Western Australia, Mt Isa in Queensland and Central Australia.

Now, as ALPA themselves acknowledge, it is apparent that, on the surface at least, the training scheme has failed to live up to expectations. None of the ALPA stores has yet come under Aboriginal management, and the level of Aboriginal employment at supervisory levels in the stores is still low. Head office in Darwin has only recently recruited an Aboriginal staff member in the general adminstrative area. There is no simple explanation for this. Possibly the training schemes, now operating on-the-job rather than through the Galiwin'ku training school, were less appropriate than previously thought. It seems that, as far as management training is concerned, this is part of the reason for failure. People aspiring to be store managers require wider experience than they can get through working solely in association with the community store: occasional visits to Darwin to gain experience in buying and to witness the operation of ALPA head office are probably not enough. It may also be that, as Wells (1993) suggests, many Aboriginal storeworkers do not really want to become the managers. Thus, although the programs offered through the training scheme have been continually adapted to meet changing wants and needs, the ultimate aim may not be what people want. Other reasons why ALPA stores have not come under Aboriginal management include the lack of enthusiasm from non-Aboriginal managers to train themselves out of their jobs as quickly as possible. Some non-Aboriginal ALPA staff have certainly continued to see Aboriginal staff as not 'ready' for responsibility.

The fact that Aboriginal people have not become ALPA store managers has inevitably disillusioned many individuals and made them sceptical about ALPA's real intentions. Above all it has focused attention on the ambivalence of ALPA's role as a service organisation concerned with community development, or as a business enterprise. Throughout ALPA's first decade community development, through the handover of the stores to full Aboriginal control and Aboriginal management, was apparently the preferred goal. However by 1983 only two of the original stores had been transferred to local Aboriginal community council ownership. In 1985 the remaining five ALPA stores were advised that, rather than contributing a proportion of the profits to a 'take-over fund' for the completion of this handover, they should consider sticking together to keep ALPA strong. The fund was then renamed the 'community reserve fund', set aside for community projects. This remains the prime thrust of ALPA's policy.

Complaints about ALPA's prices, commonly voiced by store customers, also demonstrate the ambiguities of an organisation which tries to provide a service and also be a successful commercial enterprise. Aboriginal incomes in remote communities are on average only 60 per cent of those of non-Aboriginal people. It is hardly surprising that many people would see that lower prices for food should be the priority. People have not been very sympathetic to ALPA's response, that higher prices mean higher profits which can be used to promote community development.

A final important issue concerns independence, both from mission (Uniting Church) and government influence. Although ALPA has now operated separately from the Uniting Church for over a decade many of the non-Aboriginal people working in the organisation come from that background and church policy and practice pervades their work. Official ALPA ideals still follow those of the Uniting Church, with strong work ethics, responsibility and dependability and commercial success being stressed. Within ALPA these ideals will not be quickly discarded. Aboriginal members in the communities, who may have other priorities, may well find themselves in conflict. Government influence, on the other hand, has been much smaller than for many other Aboriginal retail organisations. This is seen as an important strength.

Anangu Winkiku Stores (AWS)

Like ALPA, AWS is an Aboriginal organisation, established in 1982 and owned by Anangu Pitjantjatjaraku (AP), the group which represents the Pitjantjatjara communities which fall within South Australia. Its members consist of the stores in three of the founder communities (Amata, Pukatja (Ernabella) and Pipalyatjara), Oodnadatta store and small stores in a number of outstations associated with Amata. AWS also assists five stores across the border in the Northern Territory, outside the AP lands – Mutitjulu, Docker River, Atitjere, Aputula and Amoonguna (Figure 7.4). These are not AWS members. Their costs are met principally through ATSIC, which engages AWS as a consultant to assist these stores in achieving more efficient management. These arrangements have been made because these stores were in financial difficulties. AWS is generally recognised as the store 'trouble-shooting' body in central Australia.

The current restriction of AWS membership to AP communities simplifies administration but is not logical, either in terms of the need for such services

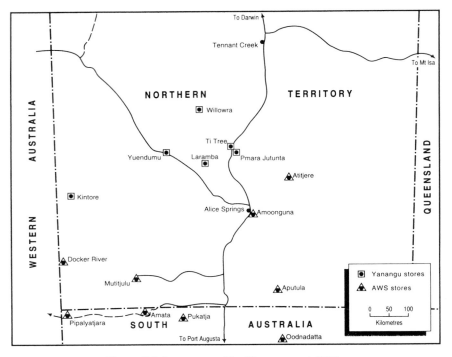

Figure 7.4 Stores served by Yanangu and AWS

or in terms of its role in supporting all Pitjantjatjara communities. AWS would prefer to be free to serve the whole of the Pitjantjatjara area – South Australia, Northern Territory and Western Australia. However it draws heavily on ATSIC funding and under the current arrangements this area falls into three ATSIC regions. This would make AWS funding applications extremely complicated. The organisation would prefer to see this bureaucratic difficulty resolved, and ultimately have a board of directors which represents all the stores with which it is involved.

AWS is a service organisation, not an enterprise. It does not own stores, and unlike ALPA bears no responsibility for the distribution of profits from stores to constituent groups. Its services include the provision of managerial advice, Aboriginal training, stock purchasing and financial services, on a fee for service basis. Each store, regardless of its size, pays the same service fee because the cost of services to AWS is not markedly affected by the size of the business. In return their use of AWS stock purchasing service alone saves them a considerable amount of money. As a full member of Independent Grocers Cooperative (IGC) in Adelaide AWS orders for goods attract service fees which are only about one-quarter to one-third of those charged to non-

IGC members. In addition, with annual purchases from IGC worth over $2 million, AWS can use its significant buying power to negotiate advantageous deals on goods in high demand in Aboriginal communities. A final advantage is that AWS, through its current and investment accounts, can back IGC's bills so that that company receives payment within the seven-day period which it stipulates. Small isolated 'bush' stores often find it very difficult to meet this commitment because they do not have the ready cash resources to pay for goods which they have not yet sold. When stores order their stock through AWS they also take advantage of a clearly structured centralised ordering system, still in the process of up-dating to accord with the latest computer technology; and co-ordinated transport arrangements, whereby AWS takes responsibility for engaging and supervising the services of the trucking company.

If communities wish, AWS will help to recruit managers for their stores. Thereafter it will provide managerial support including training. Until mid-1992 all managers for member stores have been non-Aboriginal, but Amata store has since been transferred to full Aboriginal management. This process will probably be repeated elsewhere in the future. AWS directors have adopted a resolution that the position of non-Aboriginal store manager is redundant. That means that current white managers, all of whom are on contract, may be asked to leave or, if the community wishes, their contracts may be renewed. This is seen as another important step in transferring control to the Aboriginal community.

The initial study which led to the establishment of AWS in 1982 (Green and Adamson, 1982) clearly showed that the Pitjantjatjara people wanted to run their own stores, and that they saw training for these tasks as the most important input. AWS provides on-the-job training with continual backup provided by two Alice Springs-based training officers, one working with the South Australian communities and one with the Northern Territory group. These officers spend about three weeks with their group every month, varying the time spent in each place according to need. All AWS non-Aboriginal managers must be committed to training and their contracts explicitly require them to operate in ways that are compatible with Pitjantjatjara cultural concepts and social structures.

The AWS training system is strongly modelled on that of ALPA, and the official agreement between the two groups for the use of ALPA manuals in AWS cements that connection. However AWS modules, although of similar structure to those of ALPA (including ten components, three of which are compulsory (store hygiene, store workplace, checkout etc.), and the remainder of which can be done according to ambition) have been tailored to meet local needs. AWS has also been negotiating the formalisation of its training modules through the Northern Territory Open College (NTOC), so that Aboriginal storeworkers will receive more widely recognised accreditation. No agreement has yet been reached, and in the interim AWS, like ALPA, has applied for recognition as a 'Private Provider of Adult Education' through TAFE.

As more AWS stores come under Aboriginal management the training issue will obviously become even more important. While Amata is as yet the only AWS member in this situation three of the associated stores (Apatula, Atitjere, Amoonguna) also have Aboriginal managers. All of these are small. The ATSIC consultancy fee which AWS collects for five of these stores ($15,000 for each) will in effect pay the salary of another training officer. This may be the way things go in the future, with AWS saving money on salaries for non-Aboriginal management staff, but using that to pay for better overall support through training.

AWS is essentially a non-profit service organisation and does not return a percentage of its profits to its member communities. It depends very heavily on government funding, currently from ATSIC (for the administrative part of the business) and DEET (for the training component). These funds subsidise the cost of AWS services to member stores. However it does mean that AWS regularly has to apply for funding on an annual basis, and that it must use its funds for those purposes specified in the budget. This affects its operational flexibility, and the organisation would find it much easier to plan for the future if it received block funding. AWS sees this dependence on government monies as a disadvantage, but also acknowledges that independence, such as that upheld by ALPA, comes at a cost to the organisation, to its member stores and ultimately to the Aboriginal customers of these stores. If AWS aimed for independence from government funding it would mean rethinking the whole financial situation of the organisation. Its services would no longer be subsidised, member stores would have to pay more, and AWS itself would probably be pressured into generating additional income as a commercial enterprise. This might well detract from the quality of the services it is able to provide.

The issue of Aboriginal control within the organisation has been very important from the beginning. Control was extremely strong in the initial years, but difficulties in dealing with training and financial operations led to more non-Aboriginal staff being employed, particularly at head office. The appointment of a board of directors and committee is seen as a very positive step. Aboriginal control has also been promoted through deliberately channelling AWS's communications with its members through the councils or store organisations, not through community advisers. This aims to ensure that decisions and information are made by Aboriginal people rather than non-Aboriginal administrators. However, as community adviser roles in the AP lands are increasingly being assumed by local Aborigines, this approach may be less relevant. Finally, the Aboriginal management appointment at Amata is an important step towards Aboriginal control. With the right kind of backup there seem to be no reasons why more Aboriginal managers should not be working in AWS stores.

Yanangu Stores

Yanangu Stores is an Alice Springs-based Aboriginal company, established in the mid-1980s with shares purchased by six central Australian community stores – Yuendumu, Ti Tree, Pmara Jutunta (Six Mile), Laramba (Napperby), Kintore and Willowra (Figure 7.4). Its role is as a wholesale distribution company, concerned solely with variety goods, and, as one of the largest operators of this type in Alice Springs, it can hopefully offer considerable savings to its customers, most of whom are very small-scale operators. In addition to its six member stores it serves about fifty other outlets in the region, stretching as far north as Tennant Creek, south to Indulkuna, west to Warburton and over to Ammaroo and beyond in the east. These customers are both Aboriginal and non-Aboriginal. Yanangu's directors are drawn only from the six original contributing members, although other Aboriginal store groups have expressed interest in buying shares.

Yanangu's wholesaling operation is located in an Alice Springs warehouse, and covers a wide variety of goods including clothing, camping equipment, tools and electrical appliances. Orders come by phone, through periodic visits made by Yanangu staff to outlying communities, or through customers coming to the warehouse while in town. Customers arrange for their own transport of orders, either with their own vehicles or through local freight operators.

Although Yanangu's current function lies primarily in wholesaling, it has in the past also provided managerial support. This has included employing store managers and subsequently advising them and monitoring their performance. The decision to abandon this part of the service in favour of concentration on wholesaling was made by the directors. Indirectly Yanangu staff do still advise their customers when problems are brought to their attention, but this usually takes the form of passing on information about the appropriate agency through which to seek assistance rather than offering assistance themselves. If stores need funding they are advised to approach ATSIC and DEET and if they want training courses they are asked to contact AWS. All organisations concerned with Aboriginal store support in the Alice Springs region – AWS, Yanangu, ATSIC and DEET – see the co-ordination of their efforts as very important.

Yanangu, as a wholesaling business, is primarily an enterprise, dependent on its financial viability for survival. Its current annual turnover is about $2.4 million and its profits potentially provide cash which can contribute to development in each of the member communities. Profits are not always used in this way. In the last financial year the members decided that the future financial viability of Yanangu was more important and allocated their shares of the profits for payment of the mortgage. ATSIC is not directly concerned in funding Yanangu, although the company is still paying off a government loan which was allocated to cover the costs of their premises. This means that government influence over Yanangu's operations is only minor. However the company is, as a wholesaler, much affected by the financial viability of its customers. Since it does not recruit managers, is not involved in management supervision and does not keep the books of its members it is vulnerable in this area. Problems can also arise if its member stores decide to buy their variety goods elsewhere, and stores taking this approach would be allocated a smaller share of Yanangu's profits. With its current wide range of non-member customers this problem is not as serious as might be expected.

Store support agencies: some fundamental issues

The provision of effective support for retail services in remote Australia and Canada must be the prime role of store support organisations. However they have an additional important agenda, that of fostering aboriginal self-determination in a practical and sustaining way. Satisfying both aims may well be very difficult. The first necessarily emphasises economic efficiency above other considerations; while in the second case, although efficient economic operation is of key importance, other elements of a social and political nature may be equally valued. Achieving a satisfactory balance depends on the emphasis placed within the agency as well as on the actors themselves. The level and type of aboriginal control is the most important variable. As I have discussed elsewhere (Young, 1993) it seems that on the whole agencies which have stemmed from community-based organisations, such as FCNQ and AWS, are more likely to have created structures which allow for basic aboriginal control in day-to-day organisation and decision-making than agencies which originated from ideas put forward by external agencies. Agencies like ACL or ALPA combine both elements. They were established in response both to aboriginal priorities and to needs recognised by external non-aboriginal administrators; and both have been committed to increasing aboriginal control over time. In this way they differ from the community-based organisations because they explicitly include a process of 'training for development'.

An alternative solution: stores as individual enterprises

Stores could, of course, be run not as community co-operative enterprises but as small businesses, owned and operated by individuals. This would bring into play all the attributes of the free market including competition, the fostering of individual as compared with group benefit and a more conventional entrepreneurial approach. This approach might well help to overcome problems such as low profit levels, the distribution of profits among a large number of people, and limited understanding of and commitment to the enterprise from both staff and store owners. It also accords with the concept of enterprise favoured in the wider Australian and

Canadian business communities. Successful aboriginal entrepreneurial activities undoubtedly exist, particularly in urban communities (see Byrnes, 1988; 1990). Evidence for their existence, for reasons discussed by Ellanna *et al.* (1988), is less convincing in remote communities. It is nevertheless worthwhile examining examples of individual aboriginal experiences with small retail businesses to gauge how this alternative approach might or might not contribute successfully to development.

Australia

The tradition of Aboriginal store development in Australia, based on earlier large-scale businesses run by government or mission agencies and transferred to Aboriginal ownership through active promotion of their role as community enterprises, runs counter to the idea of small family-owned retail ventures. It has also been generally assumed that such ventures would not be appropriate because of the extended family nature of Aboriginal kinship systems. As people become more familiar with non-Aboriginal approaches to business small enterprises may well become more common but until now there have been few examples of such ventures.

Ernabella, a Pitjantjatjara community near the South Australian–Northern Territory border, was established under Presbyterian Church auspices in 1936. The policies of that denomination, particularly those promoting a strongly entrenched work ethic and the value of individual success, appear to have influenced an experiment in deliberately fostering small individually owned stores to replace the large community store when that enterprise was transferred to the community in the early 1980s. Instead of continuing with the existing centralised structure, the store was progressively split into several units, each of which was to be owned and managed by a separate Aboriginal family. The butchery and bakery were first, and further plans included a delicatessen and a clothing/variety store. The original supermarket would retain only basic foodstuff retailing. People were generally enthusiastic about the idea, seeing the reduction from large- to small-scale business as one which would offer them a much better chance of coping. However problems soon arose. The baker, for example, was a very important tribal elder in the community and was often absent because of his ceremonial responsibilities. When that happened there was no bread. Not surprisingly customers complained and the supermarket was once more forced to stock its freezers with loaves from Alice Springs. The basic problem was that social responsibilities were here in conflict with the concept of commercial profit and that of providing the best possible service to customers. This has also occurred elsewhere. In an Arnhem Land outstation in the late 1970s, for example, arguments and political battles over the control of the store meant that basic services were almost non-existent. Here, as Bagshaw (1982) points out, the conflicts related to control over cash resources, not for conventional reasons of materialistic accumulation but for the status acquired when these were redistributed according to traditional systems of reciprocity. As these two examples suggest, the principles of individual entrepreneurial activity may not sit well with those that govern in remote Aboriginal communities.

Canada

Examples of individually owned aboriginal retail businesses in remote Canadian communities have also been relatively uncommon, for very similar reasons. However increasing interests in such entrepreneurial efforts have been apparent, particularly as Indian and Inuit gain higher levels of education and more experience of working elsewhere. The integration of the individual approach with community concerns (essentially an attempt to combine commercial and social interests in business operation) is of particular interest, as the example of Paddle Prairie, discussed by Robinson and Ghostkeeper (1987 and 1988) shows.

Paddle Prairie is a Metis community of between 600 and 1,000 people located in the boreal forest area of northern Alberta. It and seven neighbouring Metis settlements lie on their own land, held under fee simple title from the provincial government since 1939. Although Paddle Prairie people have had a wealth of experience both in the conventional education system (including tertiary education) and in the workforce, and are far travelled inside and outside the province, this is still a socially coherent community where attachment to family and land remain extremely important. It is also isolated. The small townships of High Level and Manning, some 70 km to the north and 130 km to the south respectively, offer a variety of basic services but for other needs Paddle Prairie people must travel much further to Peace River or further south to Edmonton. Until 1971 Paddle Prairie had its own small retail store, run under government auspices with very limited consultation with the community on the types of products and services to be offered. In that year it closed because of lack of support from the community and Paddle Prairie people were then forced to make these long journeys to satisfy even their most basic needs.

In the early 1980s a local Metis farmer, highly educated and experienced in working in a number of provincial-wide organisations involved with Metis development, decided to open a new store. His main concern was that this store, unlike its failed predecessor, would take careful account of people's needs and aspirations. While aiming at commercial profitability and good service, it would be a truly Metis store, run in the Metis way.

Although the project was formulated and implemented by the entrepreneur, from the beginning it came under the Paddle Prairie Mall Corporation, established in 1985 to raise the necessary finance and allow for community shareholding. While, because of lack of cash reserves and concern about whether the venture would succeed, people were not eager to commit their funds, they still maintained interest and could subsequently become more heavily involved in the running of the store if they wished. External investment, including funds from NEDP and provincial regional expansion funds were used to pay for store construction and to get the project going. Ghostkeeper's Store, which took the name of the entrepreneur, finally opened in late 1986 with a self-service supermarket and a fuel agency with Husky Oil. It subsequently added video rentals, a laundromat and a small tipi manufacturing business. Staff training was a priority from the beginning, as was maintaining strict control over credit. Credit was seen by the community as an iniquitous system which, as Robinson and Ghostkeeper (1988) comment, had been improperly dealt with by the government store which allowed people to borrow beyond their means. After a year's operation the store had paid off initial costs and was able to issue dividends to investors. Two years later, in 1989, success had been maintained and considerable achievements in staff training had allowed the entrepreneur to stand back from the day-to-day operations of the shop. Although still in charge of the operation he was devoting more and more time to another enterprise, his family farm which was becoming increasingly profitable as a specialised breeding farm for Morgan horses, much sought after by the Canadian Police Forces.

Aboriginal people as shopkeepers?

As these examples suggest, there may be some scope for individual retail enterprises to develop in aboriginal communities in remote areas. However such enterprises are unlikely to succeed if their operations discount cultural components in aboriginal lifestyles which are distinct from the materialistic and entrepreneurial attitudes prevalent in non-aboriginal society. In addition, success is going to depend not only on how the community views the enterprise, but also on the skills of the individual entrepreneur. This was clearly a vital factor in the establishment and on-going success of Paddle Prairie. Other problems concerning these individual businesses are that they are usually going to be small-scale operations, and hence are less likely to be able to make some of the cost savings so important in isolated communities with poor communications.

8

ABORIGINAL DEVELOPMENT IN REMOTE AREAS
Problems and prospects

Many people today would see Canada and Australia as among the most congenial and interesting countries in which to live. Both offer a wide range of economic opportunities, reflecting the richness and diversity of their resource bases; both are democratic in their politics and wholly support the principles of freedom of expression; and both, arguably, have not only embraced the concept of multiculturalism but are making it work. Yet along with these great achievements both Canada and Australia also exemplify the legacy of a colonial history which condoned the dispossession of the original inhabitants on the premise that these 'first peoples' did not use the land and its resources productively and therefore could not further 'development'. As that implies the incomers saw aboriginal Canadians and Australians as inferior. That perception still to a large extent remains. It has separated the aboriginal peoples from all others, marginalising them as a people whose needs and aspirations have long been discounted and effectively barring them from participation in the modernisation process. Ironically, although both Canada and Australia now wholly espouse the multicultural ideal, the marginalisation of Indian, Inuit and Aboriginal peoples has persisted and has only in very recent times begun to be broken down. Effectively this means that, as the title of this study suggests, the 'first peoples' of Canada and Australia have formed a 'third world in the first'. They are culturally and economically separate, with demographic and socio-economic characteristics more similar to those of many developing nations than to the industrialised countries to which they belong.

Arguably the separation between aboriginal and non-aboriginal people has been more obvious in the remote areas, Canada's vast northern lands of the Arctic and Australia's deserts and northern wetlands. Today in these regions, the most affected by positive changes associated with aboriginal land rights, marginalisation of the 'first peoples' is no longer either acceptable or practical. Here, where they form very high proportions of the population, their new controls over land and resources and their increasing political clout presents the opportunity not just to take part in development but to change its character to accord more closely with their desires for the future. In so doing the two basic questions posed in this study must be considered:

1   Why, after years of commitment to aboriginal development, have there been so few real successes and so many failures?
2   What alternative avenues for remote area aboriginal development might offer a better chance of success?

Many Canadians and Australians think that responding to these questions is straightforward. Typically simplistic answers to the first query include 'Aboriginal people will always fail – they are feckless, lazy and happy to be dependent on government hand-outs'; and, to the second, 'Nothing will work. They must be pushed into joining the mainstream.' However I contend that there are no simple answers. Development through the use of the land and its renewable resources, the mining industry, tourism and conservation uses which exploit the environmental attractions of these regions for 'city dwellers' and the provision of basic local services such as retailing is affected by a wide range of environmental and human factors. The tangled web which these create makes explanation of development problems extremely complex. Nevertheless despite that complexity important common themes emerge. These are apparent not only, as one would expect, within the Australian and Canadian contexts where there are certain historical and political uniformities. They also appear in comparisons made between these two countries. Jull's (1992b: 37) comment that 'The basic elements of contact between majority and minority, European and indigenous, centre and margin, city and country, government and people's movement are breathtakingly familiar across borders, languages, continents and cultures', supports my view that such comparison is a useful exercise.

WHY HAS ABORIGINAL DEVELOPMENT SO OFTEN FAILED?

Cross-cultural barriers

Common themes of relevance to explaining development processes and the problems created by their impact in remote Canada and Australia include the cross-cultural barriers which have been major inhibitors to aboriginal participation. These have ultimately been responsible for the misguided and inappropriate policies adopted and implemented by federal and state/ provincial/territory governments. The ethnocentrism of those in positions of political power has ensured that these policies remained in place even when their failure should have been clearly recognised. Thus assimilation was not only seen as the appropriate development approach in the past, but also persisted up until the present day, nearly thirty years after it was officially discarded.

Inflexibility of the bureaucracy

Another theme is the immobility of existing bureaucratic structures, which persists not only because people believe in it but also simply because it is there. As Porter *et al.* (1991: Ch. 7) have commented in relation to their case-study of an Australian Aid project in Kenya, development often takes on a life of its own and it is extremely difficult for practitioners to change the parameters of the process in its implementation phase, simply because of the time and money already invested in determining these. Thus, once a project receives the green light it will go ahead even when subsequent events suggest that the initial plans were based on incorrect or incomplete data and assumptions. This inflexibility destroys opportunities for fine-tuning the process to bring it more in line with aboriginal needs.

Socio-economic disadvantage

The disadvantaged socio-economic situation of aboriginal Canadians and Australians, coupled with their lack of access to or control over resources, has also inhibited the successful amalgamation of the different components of development. Particular problems have been their lack of education, of financial backing and of infrastructural support.

Legacies of the past

Another important reason why aboriginal people in remote Canada and Australia still fail to reap the benefits of development is the continuing effect of past legacies. The history of their dispossession by non-aboriginal graziers, traders, miners, missionaries and government officials has laid a foundation which still structures contemporary society. For many aboriginal people that dispossession has not only hindered them but has made their participation in the opportunities offered by development well-nigh impossible. For non-aboriginal people that history has coloured the whole perception of what development is about, creating the belief that the choices which they, as members of an industrialised, capitalistic society make are the only logical and acceptable ones. In the resultant impasse, it is the aboriginal people who have generally been the losers. In Canada and Australia, where the contrast between the industrialised world of southern cities and the world of remote-dwelling Inuit, Indians and Aborigines still determined to live according to the 'laws' of their relationship to the land is particularly clear, this is perhaps to be expected. The fact that the 'first people' of a developing country such as Botswana are also marginalised in this way provides food for thought. Here the colonial attitudes are not solely the legacy of British occupation but go further back to the earlier settlement of Basarwa country by black African pastoralists. It suggests that cross-cultural misunderstanding between hunter-gatherers and those who use the land for cultivation or grazing is very deep seated.

THE SUSTAINABLE DEVELOPMENT MODEL

The inappropriate nature of the development models favoured by the majority population in industrialised societies like Australia and Canada is another prime reason why aboriginal people are largely excluded from the process. Here the concept that development is synonymous with economic growth is the main stumbling block. The broadly defined sustainable development model, combining environmental, economic and social components, is now worth considering in greater detail. As discussed earlier (Chapter 1, Figure 1.1) this model can be represented as a simple system within which these three elements combine and overlap to form a holistic framework (Figure 8.1a). Other components of Figure 8.1 show how the main types of remote area development covered in this study – subsistence, trapping, pastoralism, mining, tourism and retailing – are placed within the framework of the model. Comparison of where each of these activities is located raises some interesting pointers.

Subsistence hunting and gathering activities fit into the core of the model, where environmental, economic and social components overlap each other (Figure 8.1b). For such activities environmental elements are fundamental – what the land and its natural resources have to offer and the importance of using these so as to ensure their future sustainability. Economic elements, also fundamental, are primarily non-monetary but subsistence can enter the monetary sphere when products obtained through such activities are further processed to become cash earning. The monetary economic element also makes an increasingly important contribution to subsistence operation because it enables people to use new technology such as vehicles, outboard motors, rifles and fishing lines. Social components are represented by factors such as traditional land ownership and the laws governing this; and by the all-embracing social networks which bind people together so that they can survive off the fragile resources available to them in environments such as the Australian desert or the Canadian Arctic.

Trapping and pastoralism, although on the surface very different types of activity, occupy similar positions in the model (Figure 8.1c). Here, although they occupy the central core they also extend into that region where only economic and environmental aspects overlap. Individual projects could sit anywhere within the region indicated. Thus, with pastoralism, keeping small 'killer' herds for community subsistence use would fall into the core (marked A); running a completely commercial operation where there were no concessions such as free meat, and no account taken of social parameters such as traditional land ownership would fall into part B; and operations which combined these two scenarios, probably the most common type, would have

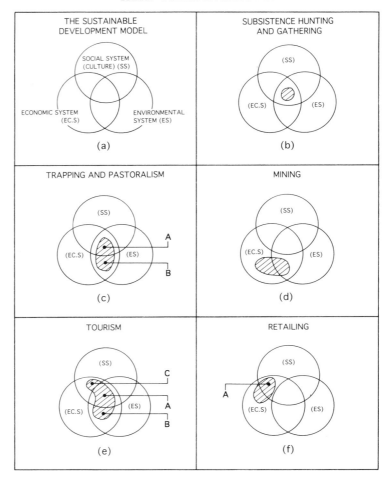

Figure 8.1 The application of the sustainable development model to aboriginal development in remote Canada and Australia

a foot in both camps, partly in A and partly in B. Trapping projects similarly would range across the spectrum indicated in the diagram, with some focusing primarily on production for family/local consumption and others aiming principally at external markets.

Mining projects, as Figure 8.1d shows, fall either into the wholly economic sphere or into that area where economic and environmental elements overlap. It would be highly unlikely that modern mining operations would take account of social elements, such as allowing the structure of a kinship network to dictate their work processes. Today, with more rigidly applied regulations regarding the environmental impact of mining, all major Canadian and Australian projects such as the Argyle Diamond Mine or Beaufort Sea oil and gas development belong to both components. However it would not be inconceivable for small, locally owned and operated mining ventures to take social elements into account. While the time of individual prospecting and mining, such as occurred during gold rushes in both the Australian and Canadian north in the early part of this century, is now past small-scale aboriginal involvement is possible. This would, however, commonly take the form of indirect participation through providing support services such as drilling operations. These types of projects can and often would incorporate social elements such as traditional relationships into their workforce structures and profit distribution systems.

The tourist industry, and related activities such as the manufacture of arts and crafts, is more broadly based and falls into three sections of the model (A, B and C in Figure 8.1e). While large-scale resort complexes dominated by national and international hotel chains and tour companies would fall into part B (economic/environmental) or part C (economic/social), many tourist services and the arts and crafts industry also function in part A. The arts and crafts industry has both a monetary and non-monetary economic component (for example, painting for sale to galleries and painting for the performance of ceremonies), and, because it draws many of its raw materials from the bush and is firmly grounded in aboriginal/land relationships, is affected by the physical environment within which it operates. Social elements are also vitally important. Where these are ignored in the manufacture of artefacts or paintings conflicts have often arisen. Artists, for example, must take care that the story which they illustrate is one which is appropriate to them as an individual. If they use another person's story they must have permission, and must ensure above all that it is acceptable for general explanation and exhibition.

Finally, what about the provision of basic services such as retail stores? These are affected both by social and economic elements or, less commonly, are governed solely by economic considerations (Figure 8.1f). The issue of whether an aboriginal community store is a service or an enterprise arises because such shops almost invariably belong to part A, where the two elements overlap. It would be highly unlikely that an aboriginal-owned and operated store in an aboriginal community functioned solely on a profit-making basis.

The different placements of these broad types of development in the sustainable model become important when the issue of their support through government agencies is considered. That support, as has been stressed already, essentially reflects non-aboriginal views on where these activities stand. Thus, for subsistence hunting and gathering, the activity has been seen as largely irrelevant and very little support has been made available. Those schemes which have developed, such as Canada's Income Support Scheme,

265

although perceived as highly innovative have often been viewed with suspicion and have taken a vast amount of time and effort to establish. In addition, it is highly questionable whether they would ever have started if the end-product had been simply country food. Despite the growing body of evidence that country foods make a very significant contribution to human sustenance the feeling persists that the cash income derived from ancillary trapping is what really counts. In Australia efforts to gain government recognition of the value of this activity, seen as absolutely crucial by many remote dwelling aboriginal people, have so far gone nowhere.

In the cases of trapping and pastoralism those activities which fall into part B accord with the general government perception of where their contribution should lie. Government support agencies such as DIAND and ATSIC and state, territory and provincial bodies will favour assistance for projects which emphasise conventional commercial profitability while safeguarding essential aspects of the environment. Social components of trapping or of pastoralism in remote aboriginal communities are seen as hindrances, 'noise' to be ignored and eradicated rather than incorporated into the system. This means, for example, that Australian Aboriginal cattle stations which prefer to concentrate on small-scale, environmentally sensitive grazing projects which are socially compatible with the structure of the group but have limited profit-making potential may find it very difficult to receive support. Applications from those that espouse full commercialisation and firmly place the social aspirations and feelings of the community in a subordinate position are likely to be viewed much more sympathetically. Mining projects, on the other hand, would rarely face such barriers. They accord very closely with non-aboriginal/government concepts of what successful development is all about. Commercially oriented tourist ventures also would be likely to receive more sympathetic assistance than those which overtly acknowledged that social elements would affect their mode of operation. However, because aboriginal people's contributions to the tourist industry are particularly valuable in the cultural sense, there is greater sympathy and understanding of these priorities than there would be with pastoralism. Finally government agencies which support retail stores tend to judge such enterprises primarily on a commercial basis and it is only when aboriginal people control and operate their own support organisation that more compatible forms of assistance are offered.

Altogether, as Figure 8.1 shows, there is a fundamental difference between development activities which fit into aboriginal lifestyles in remote Canada and Australia and those which do not. Those which are compatible are those which take the social sphere into account, and where appropriate also show sensitivity to environmental aspects. Those which omit those spheres are not compatible. Modern-day requirements on mining companies or on pastoralists to acknowledge the importance of environmental issues have gone some way towards closing the gap between the two. Environmental impact assessment is now obligatory for mining developments and pastoralists are under increasing pressures to manage their land and their stock in ways which ensure that serious degradation problems of the past are rectified and that they do not occur in the future. However the need for social impact assessment is less firmly acknowledged and it is still possible for such operators to conduct development without taking social interests into account.

The sustainable development model has another characteristic often substantially ignored by agencies supporting remote area development. Its nature is by definition long-term, aiming at changes which will be well grounded and workable and focusing on the concept of resource sustainability. Government agencies, however, often opt for quick solutions. They say that they have little choice. If they do not fulfil their promises within their term of office they will not be re-elected. Many people do not have the patience to wait for long-term solutions to show success and government financial structures are often restricted to annual funding which is certainly not conducive to long-term planning.

The sustainable development model: the issue of diversity

As the above discussion highlights, most real life situations do not fit neatly into compartments of the sustainable development model but overlap from one section to another. This reflects the diversities which are the essence of the human environmental situation. These diversities are particularly important at individual or community level and it is worth considering their impact in greater detail.

Environmental diversity

The comparative material presented in this study draws on examples from many different geographic environments. Challenges to development posed by Canada's high Arctic and Australia's central desert are obviously very different from each other in detail. Temperature, for example, is but one aspect of this. Thus in the Canadian Arctic heating is very costly and building construction and maintenance therefore very expensive. In central Australian remote communities the cost of building conventional structures is also high but in that climate many people are content to use more basic cheaper forms of shelter. These are necessary elements in many types of development project and hence their high cost becomes an important local issue. For example a small community guesthouse in the Canadian Arctic may cost so much to build and run that it will never make a profit; but in northern Australia much more flimsy types of construction are both possible and also more popular with tourists. This means that on the whole infrastructural costs would be higher in the Canadian Arctic than in the Australian outback, and as a result the problems encountered in setting up such small tourist ventures greater. Nevertheless, despite these important differences, it is the basic similarities of these two regions – their remoteness from the developed parts of these countries, the harshness of their environments for everyday living, the poverty of their soils and of their natural resource bases – which are the prime environmental characteristics to be considered.

Social/cultural diversity: gender issues

Obviously the aboriginal populations of remote areas are not homogeneous. Although traits such as the interdependent relationship with their traditional lands and their strong kinship networks are common to all they differ in language, to some extent in culture and in the types of community in which they now live. Gender is a very important component of this diversity. Questions about the different opportunities which development offers to men and women and their different responses to these opportunities have for long been an important part of the feminist geography paradigm, as demonstrated by the work of Momsen, Townsend and others (see, for example, Momsen and Townsend (1987); Brydon and Chant (1989); Momsen (1991); and Momsen and Kinnaird (eds) (1993)). In this study gender emerges as a key issue in some telling ways. In general aboriginal women from remote communities in Canada and Australia have maintained their roles and contributions to those activities where the environmental, economic and social spheres are most strongly combined, i.e. those which conform most closely to the core of the sustainable development model. Thus their subsistence contributions, exemplified in Australia through evidence produced by the work of Meehan (1982) and Devitt (1988) and in Canada through Smith's (1986) study, are very substantial, and clearly any forms of external support offered in this sphere would have to be sensitive to this. Since support for subsistence has been limited, particularly in Australia, this has not yet been a major issue. But there is evidence that insufficient attention has been paid to women's roles. The fact that access to the modern technology which is a vital part of modern-day subsistence – the outboard motors, guns and snocats used in northern Canada and the ubiquitous four-wheel drive vehicles of outback Australia – is not only dominated by but usually only available to men is an obvious example of this.

The development contribution of women through the tourist and arts and crafts industries is also becoming increasingly important. Their environmental knowledge, different from and often more detailed than that of their menfolk because of their roles as gatherers and collectors of small game, is being tapped into through their employment as park rangers and tour guides. Many women have now established their reputations as artists nationally and in some cases internationally. In Australia their increasing prominence in the fine art market is particularly obvious. Twenty years ago most established Aboriginal artists were men. Today major exhibitions such as that at the National Gallery in Canberra contain many works by women. As some of these works show, Aboriginal women artists have brought some interesting innovations into this field. In particular the production of large works of art as group projects, in which the actual painting is carried out by a number of women related to each other as classificatory mothers, sisters and daughters who together are responsible for the preservation of the stories and tradition which they are depicting, has been particularly interesting. Here the painting is produced in a way that accords closely with traditional practice, with artists sharing the work. Not surprisingly many art critics and dealers, accustomed only to works attributed to single artists, have had some difficulty in accepting this approach.

Women's involvement in other components of remote area development has been less prominent. While some of them used to work in the pastoral industry as camp cooks and occasionally as stockworkers opportunities in this field today are more limited. With the increasing technology and declining labour needs of the pastoral industry demand is unlikely to pick up. Their involvement in the trapping industry depends partly on the demand for the skins of the smaller game but here they play very important roles in preparing pelts for sale. They have had little involvement in the mining industry. However since aboriginal people have generally only been offered such jobs as tree planting their chances of getting work are probably as high as those of the men. This would not be the case with employment in heavy machinery driving. Service industries, however, do offer significant work opportunities to women. In schools, health clinics and community offices and in the retail stores discussed in this study many jobs have been perceived by the aboriginal communities as women's rather than men's work. Women accordingly have gained many skills which are also transferable elsewhere. Nevertheless, as their experiences in retailing show, their chances of going to the top and becoming store managers are not so good. Thus they become highly skilled checkout operators and book-keeping clerks but the job of bossing others around is still seen to be man's work. There are a few aboriginal women who have become store managers but they often have to face considerable opposition and find themselves under a lot of pressure.

The above summary primarily considers the direct involvement of aboriginal women in development, either through self or wage employment. This is not the only area to examine. Indirectly many of the changes affecting aboriginal communities have a particularly drastic effect on women. Increased monetary incomes, alcoholism and the social disruption caused by the absence of their menfolk are all elements which have affected women's lives, often negatively. These costs have to be taken into account along with the benefits of development. Altogether the most important thing is to realise that the role of women may well be different from but no less important than that of men and that support aimed specifically at satisfying their needs may be required.

Economic diversity

Economic components which highlight the diverse responses of remote aboriginal populations to development include their work experience, particularly the skills and attributes which they are able to offer to take up whatever opportunities arise; and their access to capital and cash to allow them to buy into opportunities. Important individual factors in these variations are age and political power and status. The influence of age is not always predictable. Thus while younger people are more adaptable and able to take advantage of training opportunities to enhance their skills, they may not have the useful skills which their predecessors had. This has applied in both the trapping and pastoral industries where older people often have better environmental knowledge and have in effect already 'lived off the land' in a way that many of their school educated children have not encountered. Access to capital and cash also, particularly where it is associated with traditional control over land and its resources, often lies more with older members of the community. Powerful political leaders also exert undue influence on sources of capital and may also dominate the most important employment opportunities in the community.

Diversity, of the physical environment and of the myriad of ways in which human beings respond to and use it, obviously has to be acknowledged in seeking workable solutions to development. However models such as that of sustainable development are still valid. The general principles of considering development as an amalgamation of economic, environmental and social elements still provide a valuable framework for policy development.

Sustainable development as a workable alternative

The idea of sustainable development as a process which aims to 'meet the needs of the present without compromising the ability of future generations to meet their own needs' (WCED, 1987), and which assesses those needs from a perspective which combines environmental, economic and social elements in people's lives is highly relevant to aboriginal Canadians and Australians living in remote communities. It is now necessary to find ways of making it work in practical terms. This presents quite a challenge. The sustainable development model is more comprehensive than others examined in this context. Thus the model of the informal and formal economy (Ross and Usher, 1986) includes some components important to sustainable development (for example, stressing the contribution of both monetary and non-monetary economic activities) but does not provide the broad environmental, economic and social framework. Similarly the idea of the next economy (Hawken, 1983) illustrates the value of integrating economic development totally into the society within which it is embedded but does not explicitly see this within an environmental framework.

The practice of sustainable development would contain a number of key elements. Firstly, it would need to be approached through consultation at the grassroots community level, where people would have the chance to voice their current needs and future aspirations and where they could state how they hoped to realise these dreams. This, as Chambers (1983) describes, means 'putting the last first'. As he also points out this is not easy and would certainly require a substantial reversal in thinking for most non-aboriginal development experts and practitioners.

Secondly, in order to encompass both shorter- and longer-term require-ments, the approach to sustainable development has to be very flexible. It also, as suggested above, has to be able to cope with geographical and human diversity. Here the approach advocated by Porter *et al.* (1991) is highly relevant. They suggest that a pluralist-orientation, whereby development occurs in full understanding not only of the rich diversity of the society experiencing change but also with sufficient inbuilt flexibility to allow for adaptation as needed during the development process, provides the best chance of a workable alternative approach. They contrast this with the control-orientation approach, whereby development occurs within a frame-work dominated by conventional assessment methods, normally quantita-tive, and relies on a rigid plan devised before commencement of the process. The latter, as has been implied throughout this study, is the approach most often taken by government bureaucracies.

Porter *et al.* (1991) also distinguish three formidable barriers to applying a pluralist-orientation to development. Firstly, most forms of development are strongly controlled by 'international fashions and imperatives' – the rules enforced by the agencies responsible for providing assistance, and the predominantly monetarist policies which they promote. Here Porter *et al.* refer to the influence of organisations such as the World Bank. Equivalent bodies in the Australian and Canadian aboriginal contexts would be ATSIC and DIAND. Secondly, development is affected by 'intellectual baggage', the attitudes and belief systems which govern the thinking of individuals who work for the support agencies. These are often based on inappropriate ethnocentric views. In the Australian and Canadian contexts many of those who have worked in the relevant bureaucracies have continued to espouse assimilation after that policy was superseded by self-determination. Finally, as Porter *et al.* note 'institutional egos and individual contests' do not sit easily with the pluralist approach. Here they refer to how agencies and the individuals who work for them often compete to ensure that they retain control over their own roles and responsibilities and ensure that their status and egos are boosted in the process. Here an example is the reluctance of

271

Australian and Canadian government departments to co-ordinate their work so that they provide the best possible development package for the client. They are often too concerned with guarding their own backs. Altogether, as Porter *et al.* (1991: 212) point out, a more pluralist-oriented development approach means 'accept the uncertainty and welcome the diversity, and face up to the realities'.

Finally Elias (1991) advocates that an adversarial approach to development would greatly improve the chances of changing the attitudes of those who control the necessary resources, and would also strengthen the resolve and political acumen of the clients. As he points out the beneficiaries, in his case aboriginal Canadians, must not passively wait to find out what opportunities are being offered to them but take the initiative and force people to listen and act upon their views. As implied such a positive approach would be part of a development system directed and controlled by aboriginal people, operated by communities rather than by individuals, and holistic in nature. All of these attributes accord with a sustainable development framework.

What, then, would a practical approach to implementing a model of sustainable development for aboriginal people in remote areas look like? It would have to include the following:

1   acknowledgement of the need for aboriginal control over development, both at national and regional levels and also at community levels;
2   recognition of the social and economic diversity of aboriginal communities, and of how this affects their responses to development opportunities;
3   negotiation (a step beyond consultation) to ensure both that the development is following an appropriate path, and that those concerned get their fair share of the benefits flowing from it;
4   deliberate fostering of aboriginal skills to enhance their chances of more equal participation in development affecting them;
5   creation of an aboriginal resource base, over which they exert contol in ways which allow them to determine how these resources might be used for development;
6   provision of adequate and appropriate forms of financial backing so that aboriginal groups and individuals can offer the capital needed to carry out their plans.

Issues such as these have been raised throughout this study. It remains to pinpoint the realities of these by referring to some contemporary debates within this area. For aboriginal people in remote parts of Canada and Australia the prime focus of these debates concerns land rights and claims, control over land use and planning for its management, and access to the resources needed to implement meaningful self-determination. These needs are increasingly perceived as centring on the establishment of constitutionally recognised regional self-government. At present Nunavut, in the process of establishment as a self-governing territory covering Canada's eastern Arctic, provides the best example of such initiatives. It is a prototype being watched with great interest by Australian Aboriginal peoples from similarly remote areas.

The creation of Nunavut as a self-governing territory of Canada, as Merritt *et al.* (1989), Merritt and Fenge (1990), Jull (1991) and others have described, has long been seen by the Inuit as an essential element in their achievement of a worthwhile future for themselves and their successors. With the ratification of the land claim agreement between the Inuit and the Canadian government in late 1992 its establishment becomes a certainty. Key elements within the structure of Nunavut include not only the setting up of its political and administrative base in Iqaluit but also the prominent role which will be played by land and natural resource management. Indeed, as Holmes (1992) notes, Canadian approaches have increasingly acknowledged the essential linkages between the settlement of native land claims and the subsequent institution of measures covering land-use planning. In that process power over land allocation and its management is to be devolved from central government (Ottawa in the case of Nunavut) to regional bodies set up for that purpose. That emphasis on land management, coupled with regionally based responsibility for distribution of resources and provision of services gives Inuit, over 85 per cent of the new territory's population, a genuine basis for controlling their destiny. It also means that in Nunavut Canada is providing a practical example of the kind of process needed to achieve the ideal of sustainable aboriginal development. Nobody anticipates that this will be easy. As Jull (1992a) stresses, making Nunavut work requires changing the attitudes which dominate existing constitutional structures and also requires strength and determination on the part of the Inuit. Doubtless compromises will be necessary. As Fenge's (1993) discussion shows the Inuit ultimately failed to get the government to carry out its stated intentions of using the land claim negotiations to establish a comprehensive national park system, incorporating Inuit joint management, in the Arctic. They had seen this as a vital first step in developing co-management systems which would also operate on crown lands outside the areas granted as part of the claim. Altogether, as all involved in the Nunavut negotiations would confirm, the achievement of workable sustainable development planning is a slow process, where the gains can only be measured in the long term.

Australian Aboriginal groups in remote areas have become increasingly well informed about Nunavut and, as several recent conferences have demonstrated (see, for example, Jull *et al.*, 1994), are interested in adapting the Canadian experience to meet their own situation. These interests have since 1992 been linked to the Mabo High Court decision (see Chapter 2) and Aboriginal self-government is now being discussed as a vital element in the national reconciliation process. The issue of regional development, including control over land allocation and management, is particularly important in the

273

Kimberley region of Western Australia. As Peter Yu, Director of the Kimberley Land Council (KLC), has stressed Aboriginal people will never be able to break off the shackles of the colonial past until they control their own destiny through taking responsibility for the resources needed to support the people in their region. A recent study of the Kimberley Aboriginal economy commissioned by the KLC (Crough and Christopherson, 1993b) has shown that the financial resources granted to the Kimberley for support of Aboriginal communities accounted for approximately 40 per cent of the region's GDP. On that basis, it is argued, the Aborigines are not only the most important component of the permanent population of the Kimberley region but are also the backbone of its economy. The KLC, in conjunction with other Kimberley Aboriginal organisations, are now asking for negotiations with the federal government so that funds allocated for their health and education services, for the support of their enterprises and for other vital components of their communities be handed over to a designated Kimberley Aboriginal umbrella organisation for administration and distribution. They see regional land allocation and management as a vital component of this. The existing system, whereby most Aboriginal government funding comes either directly or indirectly through ATSIC, is perceived not only to be working inefficiently but also to be inappropriate because the government is still the 'boss' within that organisation. Thus the kind of structure under discussion would do two important things: it would hand control and responsibility to Aboriginal grassroots organisations rather than to an organisation established 'top-down' under government auspices; and it would shift decision-making for Kimberley affairs from Canberra and Perth to their rightful place, in Derby, Broome or Kununurra.

Ideas such as these are a vital part of making sustainable development work for aboriginal groups. However once accepted in principle they also have to be made to work on the ground. Here it is necessary to examine what this type of development process means for individuals and communities. This problem forms the basis of on-going research in Australia into sustainable land management and Aboriginal community development. This study (Young and Ross, 1993) is investigating at the more detailed level of single or grouped communities what kind of planning process is involved, and the outcomes of that process. Several Aboriginal groups, including some from the Kimberley, along with others from central Australia, the south coast of New South Wales and the northern fringes of South Australia are participating in this study. Even at this preliminary stage they are producing plans which cover issues ranging from pastoral development, to the establishment of outstations on traditional lands returned to them following land claims, to achieving input into the management of parks on their land, to schemes to improve community life for their young people and projects designed to counteract the social effects of high unemployment and marginalisation.

Land and resource issues are central to all their ideas, and control over land is in every case the primary priority. Such outcomes confirm the validity of the idea of sustainable development and, hopefully, will raise the awareness of others to find practical ways of making it work.

REFERENCES

Abele, F. (1987) 'Canadian contradictions: forty years of northern political development', *Arctic* 40, 4: 310–20.
—— (1989) *Gathering Strength*, Komatik Series No. 1, Arctic Institute of North America, Calgary: University of Calgary.
ADC (Aboriginal Development Commission) (1981) *Annual Report 1980–1981*, Canberra: AGPS.
ADC (1986) *Annual Report 1985–1986*, Canberra: AGPS.
ADC (1988) *Corporate Plan*, Canberra: AGPS.
ADC (1990) *Annual Report, 1988/89*, Canberra: AGPS.
Altman, J.C. (1982) 'Maningrida outstations: a preliminary economic overview', in E.A. Young and E.K. Fisk (eds), *Small Rural Communities*, Vol. 3, The Aboriginal Component in the Australian Economy, Canberra: Development Studies Centre, Australian National University.
—— (1985) *Report on the Review of the Aboriginal Benefit Trust Account (and Related Financial Matters) in the Northern Territory Land Rights Legislation*, Canberra: AGPS.
—— (1987) *Hunter-Gatherers Today; an Aboriginal Economy in North Australia*, Canberra: AIAS.
—— (1988) *Aborigines, Tourism, and Development: the Northern Territory Experience*, Darwin: NARU, ANU.
—— (Chairman) (1989) *The Aboriginal Arts and Crafts Industry*, Report of the Review Committee, Canberra: AGPS.
—— (1990) 'Selling Aboriginal art', in J. Altman and L. Taylor (eds), *Marketing Aboriginal Art in the 1990s*, Canberra: Aboriginal Studies Press.
Altman, J. and Taylor, L. (1987) *The Economic Viability of Aboriginal Outstations and Homelands*, Report to the Australian Council for Employment and Training, Canberra: AGPS.
Ames, R., Axford, D., Usher, P., Weick, E. and Wenzel, G. (1989) *Keeping on the Land: a Study of the Feasibility of a Comprehensive Wildlife Harvest Support Programme in the NWT*, Ottawa: CARC.
Arctic Cooperatives Ltd (ACL) (1990) *The Co-operative Movement in Northwest Territories: an Overview 1959–89*, Manitoba: Winnipeg.
Asch, M. (1982) 'Capital and economic development: a critical appraisal of the recommendations of the Mackenzie Valley Pipeline Commission', *Culture* 2, 3: 3–9.
ATSIC (Aboriginal and Torres Strait Islander Commission) (1991) *Evaluation of the Enterprise Program*, Canberra: Office of Evaluation and Audit, ATSIC.

ATSIC (1992a) *Impact Evaluation: Land Acquisition Program*, Canberra: Office of Evaluation and Audit, ATSIC.

ATSIC (1992b) *Corporate Plan, 1992–1996*, Canberra: ATSIC.

Auditor-General (1989) *Special Audit Report. The Aboriginal Development Commission and The Department of Aboriginal Affairs*, Canberra: AGPS.

Australian Institute of Aboriginal Studies (1984) *Aborigines and Uranium: Report on the Social Impact of Uranium Mining on the Aborigines of the Northern Territory*, Canberra: AGPS.

Australian Institute of Health and Welfare (1992) *Australia's Health 1992*, third biennial report of the Australian Insititute of Health and Welfare, Canberra: AGPS.

Australian National Parks and Wildlife Service (1991) *Uluru (Ayers Rock–Mount Olga) Plan of Management*, Uluru-Kata Tjuta Board of Management, Canberra: ANPWS.

Bagshaw, G. (1982) 'Whose store at Jimarda?', in P. Loveday (ed.), *Service Delivery to Outstations*, Darwin: NARU, ANU.

Barbier, E.B. (1987) 'The concept of sustainable economic development', *Environmental Conservation* 14, 2: 101–10.

Bear, L.L., Boldt, M. and Long, J.A. (eds) (1984) *Pathways to Self-determination: Canadian Indians and the Canadian State*, Toronto: University of Toronto Press.

Beaver, J. (1979) *To Have What Is One's Own*, Report from the President of the National Indian Socio-Economic Development Committee (NISEDC) to the Minister, DIAND and the President, NIB, Ottawa: NISEDC.

Berger, T.R. (1977) *Northern Frontier, Northern Homeland: the Report of the Mackenzie Valley Pipeline Inquiry, I and II*, Canada, Ottawa: Supply and Services.

Berndt, R.M. and Berndt, C.H. (1987) *End of an Era: Aboriginal Labour in the Northern Territory*, Canberra: AIAS.

Biesele, M., Guenther, M., Hitchcock, R., Lee, R. and Macgregor, J. (1989) 'Hunters, clients and squatters: the contemporary socio-economic status of Botswana's Basarwa', *African Study Monographs* 9, 3: 109–51.

Biesele, M., Hubbard, D. and Ford, J. (1991) *Land Issues in Nyae Nyae: a Communal Areas Example in Namibia*, Windhoek: NNDFN.

Blishen, B.R., Lockhart, A., Craib, P. and Lockhart, E. (1979) *Socio-economic Model for Northern Development*, Report prepared for Research Branch, Policy, Research and Evaluation Group, Ottawa: Department of Indian and Northern Affairs.

Blomstrom, M. and Hettne, B. (1984) *Development Theory in Transition*, London: Zed Books.

Bone, R.M. (1983) *Norman Wells Database Project*, Presentations at the Calgary Workshop, 7 July 1983, Report 3–83.

—— (1992) *The Geography of the Canadian North*, Oxford: Oxford University Press.

Bone, R.M. and Mahanic, R.J. (1984) 'Norman Wells: the oil center of the Northwest Territories', *Arctic* 37, 1: 53–60.

Bone, R.M. and Stewart, D.A. (1987) 'The Norman Wells oilfield expansion and pipeline project: impacts on local communities, *Polar Record* 23, 147: 714–15.

Brody, H. (1975) *The People's Land*, Harmondsworth: Penguin.

—— (1981) *Maps and Dreams: Indians and the British Columbia Frontier*, London: Jill Norman and Hobhouse.

Brookfield, H.C. (1975) *Interdependent Development*, London: Methuen.

Brydon, L. and Chant, S. (1989) *Women in the Third World: Gender Issues in Rural and Urban Areas*, London: Edward Elgar.

Buchanan, K. (1977) 'Reflections on a "dirty word"', in R. Peet (ed.), *Radical Geography: Alternative Viewpoints on Contemporary Social Issues*, London: Methuen.

Bunner, P. and Gallagher, T. (1987) 'The company of misadventurers', *Western Report*, 23/2/87: 18–22.

Burger, J. (1987) *Report from the Frontier: the State of the World's Indigenous People*, London: Zed Books.

Byrnes, J. (1988) *Enterprises in Aboriginal Australia: Fifty Case Studies*, a report of interviews conducted across Australia, particularly Central and Western Australia during 1988, Armidale: Rural Development Centre, University of New England.

—— (1990) *Aboriginal Economic Independence: a Report on Some Canadian Initiatives*, Armidale: Rural Development Centre, University of New England.

Canada Employment and Immigration Commission (1989) *Success in the Works: a Profile of Canada's Emerging Workforce*, Ottawa: CEIC.

Canadian Arctic Resources Committee (CARC) (1988) *Changing Times, Challenging Agendas: Economic and Political Issues in Canada's North*, Ottawa: Canadian Arctic Resources Committee.

Canadian Ministry of Supply and Services (1989) *The Canadian Aboriginal Economic Development Strategy*, Ottawa.

Cane, S. and Stanley, O. (1985) *Land Use and Resources in Desert Homelands*, Darwin: NARU, ANU.

Cant, G. (1993) 'Windows into process: the context of indigenous land rights', in G. Cant, J. Overton and E. Pawson (eds), *Indigenous Land Rights in Commonwealth Countries: Dispossession, Negotiation and Community Action*, Christchurch: University of Canterbury.

Carrad, B., Lea, D.A.M. and Talyaga, K.K. (1982) *Enga: Foundations for Development*, Armidale: University of New England.

Chamberlin, J.E. (1983) 'Gathering and governing: renewable resources and the economy of the north', unpublished paper, Department of English, University of Toronto.

Chambers, R. (1983) *Rural Development: Putting the Last First*, London: Longman.

Chatwin, B. (1987) *The Songlines*, London: Picador.

Childers, G., Stanley, J. and Rick, K. (1982) *Government Settlement or People's Community: a Study of Local Institutions in Ghanzi District*, MLGL, Gabarone: University of Wisconsin-Madison.

Christensen, W. (1985) *Aborigines and the Argyle Diamond Project. Submission to the Aboriginal Land Inquiry*, Canberra: East Kimberley Working Paper No. 3, CRES, ANU.

Clad, J. (1984) 'Conservation and indigenous peoples', *Cultural Survival Quarterly* 8, 4: 68–73.

Coates, K. (1988) 'The federal government and native communities in the Yukon Territory, 1945 to 1973', in G. Dacks and K. Coates (eds), *Northern Communities: the Prospects for Empowerment*, Edmonton: Boreal Institute, University of Alberta.

Commonwealth Capital Fund (1969–74) *Annual Reports, 1969 to 1974*, Canberra: AGPS.

Connell, J.C. (1991) 'Compensation and conflict: the Bougainville copper mine, Papua New Guinea', in J. Connell and R. Howitt (eds), *Mining and Indigenous People in Australasia*, Sydney: University of Sydney Press.

Coombs, H.C. (1978) *Kulinma: Listening to Aborigines*, Canberra: ANU Press.

Coombs, H.C., McCann, H., Ross, H. and Williams, N.M. (eds) (1989) *Land of Promises: Aborigines and Development in the East Kimberley*, Canberra: CRES/AIAS.

Courtenay, P.P. (1982) *Northern Australia*, Melbourne: Longman Cheshire.

Cousins, D. (1985) 'Aboriginal employment in the mining industry', in P. Loveday and D. Wade-Marshall (eds), *Economy and People in the North*, Darwin: NARU, ANU.

Cowlishaw, G. (1983) 'Blackfella boss: a study of a Northern Territory cattle station', *Social Analysis* 13: 54–69.

Crittenden, R. and Lea, D. (1989) 'Whose wants and needs in "needs based" planning? Some examples from the provincial integrated rural development programmes in Papua New Guinea', *Public Administration and Development* 9: 471–86.

Crough, G. (1993) *Visible and Invisible: Aboriginal People in the Economy of Northern Australia*, Darwin: NARU, ANU.

Crough, G. and Christopherson, C. (1993a) 'Some perspectives on the Arnhem Land Progress Association', in E. Young, G. Crough and C. Christopherson, *An Evaluation of Store Enterprises in Aboriginal Communities*, Darwin: NARU, ANU.

—— (1993b) *Aboriginal People in the Economy of the Kimberley Region*, Darwin: NARU, ANU.

Crough, G., Howitt, R. and Pritchard, B. (1989) *Aboriginal Economic Development in Central Australia*, Report for the combined Aboriginal organisations of Alice Springs, Alice Springs: Combined Aboriginal Organisations of Alice Springs.

DAA (Department of Aboriginal Affairs) (1975) *Annual Report, 1974–1975*, Canberra: AGPS.

Dacks, G. (1981) *A Choice of Futures: Politics in the Canadian North*, Toronto: Methuen.

—— (1983) 'Worker-controlled native enterprises: a vehicle for community development in northern Canada?', *The Canadian Journal of Native Studies* 3, 2: 289–310.

Dagmar, H. (1982) 'Marginal Australians: a prelude to political involvement', in M.C. Howard (ed.), *Whitefella Business*, Philadelphia: Institute for the Study of Human Issues.

Dahl, J. (1982) 'Mining and local communities: a short comparison of mining in the Eastern Canadian Arctic (Nanisivik/Arctic Bay) and Greenland (Marmorilik/Uummannaq)', *Inuit Studies* 6, 1: 145–57.

Dale, A. (1992) 'Aboriginal councils and natural resource use planning: participation by bargaining and negotiation', *Australian Geographical Studies* 30, 1: 9–26.

Data, K. and Murray, A. (1989) 'The rights of minorities and subject peoples in Botswana: a historical evaluation', in J. Holm and P. Molutsi (eds), *Democracy in Botswana*, Proceedings from a Symposium 1–5 August 1988, Macmillan Botswana: Botswana Society, University of Botswana.

Davies, P. (1984) *Shetah Drilling Ltd. Arctic Petroleum Operators Association Review, Special Report* 7, 1: 11–15.

Dene Nation (1984) *Statement of the Purpose, Objectives, Approach and Work Program of the Task Force on Dene Economic Development*, NWT.

Department of Regional and Industrial Expansion (1984) *The Native Economic Development Program Proposal Development Guide*, Canada: DRIE.

Department of Regional and Industrial Expansion (1988) *Native Economic Programs: Review of NEDP Element I, Final Report*, Ottawa: DRIE.

Deputy Prime Minister's Office (1986) *Improved Program Delivery: Indians and*

Natives: a Study Team Report to the Task Force on Program Review, April 1985, Ottawa: DPMO.

Devine, M. (1992) 'The Dene nation: coming full circle', *Arctic Circle* 2, 5: 12–19.

Devitt, J. (1988) 'Contemporary Aboriginal women and subsistence in remote arid Australia', unpublished Ph.D. thesis, Department of Anthropology and Sociology, University of Queensland.

DIAND (Department of Indian and Northern Development) (1980) *Indian Conditions: a Survey*, Ottawa: DIAND.

DIAND (1981) *In All Fairness: a Native Claims Policy*, Comprehensive Claims, Ottawa: DIAND.

DIAND (1982) *Outstanding Business: a Native Claims Policy*, Specific Claims, Ottawa: DIAND.

DIAND (1984a) *The Western Arctic Claim: the Inuvialuit Final Agreement*, Ottawa: DIAND.

DIAND (1984b) 'A study of northern food costs – statistical references', unpublished draft, Ottawa: Economic Strategy Division, DIAND.

DIAND (1985) *Task Force on Indian Economic Development: Report to the Minister, Indian and Northern Affairs, 16 Dec. 1985*, Ottawa: DIAND.

DIAND (1986) *Task Force on Indian Economic Development: Summary of the Report to the Deputy Minister, Indian and Northern Affairs*, Ottawa: DIAND.

DIAND (1988a) *Dene/Metis Comprehensive Land Claim Agreement in Principle*, Ottawa: DIAND.

DIAND (1988b) *Western Arctic (Inuvialuit) Claim Implementation, Annual Report 87–88*, Ottawa: DIAND.

DIAND (1989a) *Council for Yukon Indians Comprehensive Land Claims in Principle*, Ottawa: DIAND.

DIAND (1989b) *Estimates for 1988–89: Part III, the Expenditure Plan*, Ottawa: Ministry of Supply and Services.

Dixon, R.A. and Dillon, M.C. (eds) (1990) *Aborigines and Diamond Mining*, Perth: University of Western Australia Press.

Dome Petroleum Ltd (1982) 'Northern participation: employment and training', unpublished report, Dome Petroleum, Calgary.

DPA Consulting Group Inc. (1985) *Program Evaluation of the Indian Economic Development Fund*, Consultancy Report prepared for DIAND, Ottawa: DIAND.

Drakakis-Smith, D. (1980) 'Alice through the looking glass: marginalisation in the Aboriginal town camps of Alice Springs', *Environment and Planning A*, 12: 427–48.

—— (1984a) 'Underdevelopment in the tropics: the case of North Australia', *Singapore Journal of Tropical Geography* 5, 2: 125–37.

—— (1984b) 'Internal colonialism and the geographical transfer of value: an analysis of Aboriginal Australia', in D.K. Forbes and P.J. Rimmer (eds) *Uneven Development and the Geographical Transfer of Value*, Canberra: Human Geography, ANU.

Driben, P. and Trudeau, R.S. (1983) *When Freedom is Lost. The Dark Side of the Relationship between Government and the Fort Hope Band*, Toronto: University of Toronto Press.

Duerden, F. (1992) 'A critical look at sustainable development in the Canadian north', *Arctic* 45, 3: 219–25.

Durack, M. (1959) *Kings in Grass Castles*, London: Corgi.

—— (1983) *Sons in the Saddle*, London: Corgi.

Dyck, N. (ed.) (1985) *Indigenous Peoples and the Nation-State: Fourth World Politics in Canada, Australia and Norway*, Social and Economic Papers No. 14, Memorial University of Newfoundland, St Johns: University of Newfoundland.

Elias, D. (1991) *Development of Aboriginal People's Communities*, York University, Ontario: Captus Press.

Ellanna, L., Loveday, P., Stanley, O. and Young, E. (1988) *Economic Enterprises in Aboriginal Communities in the Northern Territory*, Darwin: NARU, ANU.

Erasmus, G. (1986) 'NSR comment', *Native Studies Review* 2, 2: 53–64.

ESD (Ecologically Sustainable Development) (1991) *Ecologically Sustainable Development Working Groups: Final Report – Executive Summaries*, Canberra: AGPS.

ESD (1992) *National Strategy for Ecologically Sustainable Development*, Canberra: AGPS.

Federation des Cooperatives de Nouveau Quebec (FCNQ) (1988) *Growing with Co-ops*, Montreal: FCNQ.

Feit, H.A. (1982a) 'The future of hunters within nation-states: anthropology and the James Bay Cree', in E. Leacock and R. Lee (eds), *Politics and History in Band Societies*, London: Cambridge University Press.

—— (1982b) 'The Income Security Program for Cree hunters in Quebec: an experiment in increasing the autonomy of hunters in a developed nation state', *Canadian Journal of Anthropology* 3, 1: 57–70.

Fenge, T. (1993) 'National parks in the Canadian Arctic: the case of the Nunavut land claim agreement', in G. Cant, J. Overton and E. Pawson (eds), *Indigenous Land Rights in Commonwealth Countries, Dispossession, Negotiation and Community Action*, Christchurch: University of Canterbury.

Fisk, E.K (1982) 'Development and aid in the South Pacific in the 1980s', *Australian Outlook* 36, 2: 32–7.

—— (1985) *The Aboriginal Economy in Town and Country*, Sydney: Allen & Unwin.

Forbes, D.K. (1984) *The Geography of Underdevelopment*, London: Croom Helm.

Forbes, D.K. and Rimmer, P. (1984) *Uneven Development and the Geographical Transfer of Value*, Canberra: Human Geography, ANU.

Fox, R.W., Kelleher, G.H. and Kerr, C.B. (1977) *Ranger Uranium Environmental Inquiry*, Second Report, Canberra: AGPS.

Frank, A.G. (1967) *Capitalism and Underdevelopment in Latin America*, New York: Monthly Review Press.

Freeman, M.M.R. (1976) *Inuit Land Use and Occupancy Project*, Ottawa: DIAND.

—— (1985a) 'Effects of petroleum activities on the ecology of Arctic man', in F.R. Englehardt (ed.), *Petroleum Effects in the Arctic Environment*, London and New York: Elsevier Applied Science Publishers.

—— (1985b) 'Appeal to tradition: different perspectives on Arctic wildlife management', in J. Bristed, J. Dahl, J. Gray, H.C. Gullov, G. Henriksen, J.B. Jorgensen and I. Kleivan (eds), *Native Power: the Onset for Autonomy and Nationhood of Indigenous Peoples*, Oslo: Universitetforlaget AS.

—— (1986) 'Renewable resources, economics and native communities', in *Native People and Renewable Resource Management*, 1986 Symposium of the Alberta Society of Professional Biologists, Edmonton.

Frideres, J.S. (1983) *Native People in Canada: Contemporary Conflicts*, Ontario: Prentice-Hall Canada Inc.

Friedmann, J. (1966) *Regional Development Policy*, Cambridge, Massachusetts: MIT Press.

Gale, F. (1972) *Urban Aborigines*, Canberra: ANU Press.

Gardner, J.E. and Nelson, J.G. (1981) 'National parks and native peoples in N.

Canada, Alaska and Northern Australia', *Environmental Conservation* 8, 3: 207–15.

Gibson, C. (1989) *Sectoral Profiles: NEDP III, SARDA Commercial Undertakings, SARDA Primary Producing Activities*, Report prepared for Native Economic Programs, Ottawa: ISTC.

Gillespie, D. (1990) 'Conservation and the Aboriginal community', paper delivered to the Environment 90 Conference, Darwin, 8 March 1990.

Golvan, C. (1990) 'Aboriginal art and copyright infringement', in J. Altman and L. Taylor (eds), *Marketing Aboriginal Art in the 1990s*, Canberra: Aboriginal Studies Press.

Good, K. (1992) 'Interpreting the exceptionality of Botswana', *Journal of Modern African Studies* 30, 1: 69–95.

——— (1993) 'At the ends of the ladder: radical inequalities in Botswana', *Journal of Modern African Studies* 31, 2: 203–30.

Gorman, M. (1986) 'Dene Community development: lessons from the Norman Wells project', *Alternatives* 13, 1: 10–12.

Government of Northwest Territories (GNWT) (1981) *Outpost Camp Policy.*

Goulet, D. (1973) *The Cruel Choice: a New Theory in the Concept of Development*, New York: Atheneum.

——— (1980) 'Development experts: the one-eyed giants', *World Development* 8: 481–9.

Gray, A. and Smith, L. (1983) 'The size of the Aboriginal population', *Australian Aboriginal Studies* 1, 1: 2–9.

Green, P. and Adamson, T. (1982) *Pitjantjatjara Store Cooperative Report*, Alice Springs: Pitjantjatjara Council.

Griffith, R. (1986) 'Northern park development: the case of snowdrift', *Alternatives* 13, 1: 26–30.

Guenther, M.G. (1979) *The Farm Bushmen of the Ghanzi District, Botswana*, Stuttgart: Hochschul Verlag.

Hamley, W. (1991) 'Tourism in North West Territories', *Geographical Review* 81, 4: 389–99.

Hartwig, M. (1978) 'Capitalism and Aborigines: the theory of internal colonialism and its rivals', in E. Wheelwright and K. Buckley (eds), *Essays in the Political Economy of Australian Capitalism*, Vol. 3, Sydney: ANZ Book Co.

Hawke, S. and Gallagher, M. (1989) *Noonkanbah*, Fremantle: Fremantle Arts Centre Press.

Hawken, P. (1983) *The Next Economy*, North Ryde, NSW: Angus & Robertson.

Hawthorn, H.B. (1966–1967) *A Survey of the Contemporary Indians of Canada: Economic, Political, Educational Needs and Policies*, 2 vols, Ottawa: Queens Printer.

Head, L. (1990) 'Conservation and Aboriginal land rights: when green is not black', *Australian Natural History* 23, 6: 448–54.

Hiatt, L. (ed.) (1984) *Aboriginal Landowners*, Oceania Monograph No. 27, University of Sydney, Sydney.

Hirschman, A.O. (1958) *The Strategy of Economic Development*, New Haven, Connecticut: Yale University Press.

Hitchcock, R. (1980) 'Tradition, social justice and land reform in central Botswana', *Journal of African Law* 24: 1–34.

——— (1987) 'Socioeconomic change among the Basarwa in Botswana: an ethnohistorical analysis', *Ethnohistory* 34, 3: 219–55.

——— (1992a) 'The rural population living outside recognised villages', in D. Nteta and J. Hermans (eds), *Sustainable Rural Development*, Proceedings of a

Workshop, Gabarone, 13–15 April 1992, Gabarone: The Botswana Society.
—— (1992b) *Communities and Consensus: an Evaluation of the Activities of the Nyae Nyae Farmers Cooperative and the Nyae Nyae Development Foundation in Northeastern Namibia*, Report to Ford Foundation and NNDFN, Windhoek and New York.

Hitchcock, R. and Bixler, D. (1991) 'Politics, ecology and survival tactics among the Kalahari San', unpublished paper at conference on 'Indigenous Peoples in Remote Regions: a Global Perspective', University of Victoria, British Columbia.

Hitchcock, R. and Brandenburgh, R. (1990) 'Tourism, conservation and culture in the Kalahari Desert, Botswana', *Cultural Survival Quarterly* 14, 2: 20–4.

Hitchcock, R. and Holm, J. (1991) 'Bureaucratic domination of hunter-gatherer societies: a study of the San of Botswana', unpublished seminar paper.

Hobart, C.W. (1981) 'Impacts of industrial employment on hunting and trapping among Canadian Inuit', *Proceedings of the First International Symposium on Renewable Resources and the Economy of the North*, Association of Canadian Universities for Northern Studies, Canada Man and the Biosphere Program: 202–18.

—— (1984) 'Impact of resource development projects on indigenous people', in D. Detomasi and J. Gartrell (eds), *Resource Communities: a Decade of Disruption*, Boulder, Colorado: Westview Press.

Holmes, J. (1992) *Strategic Regional Planning on the Northern frontiers*, Discussion Paper No. 4, Darwin: NARU, ANU.

House of Commons (1983) *Indian Self-government in Canada: Report of the Special Committee*, Ottawa: Queen's Printer.

House of Commons (1986) *The Fur Issue: Cultural Continuity, Economic Opportunity*, Report of the House of Commons Standing Committee on Aboriginal Affairs and Northern Development, Ottawa: Queen's Printer.

House of Representatives Standing Committee on Aboriginal Affairs (1987) *Return to Country*, Canberra: AGPS.

House of Representatives Standing Committee on Aboriginal Affairs (1988) *The Effectiveness of Support Services for Aboriginal and Torres Strait Island Communities*, Interim Report, Canberra: AGPS.

House of Representatives Standing Committee on Aboriginal Affairs (1989) *Aboriginal Affairs House of Representatives Committee Report on Skilling for Self-determination*, Report tabled in Parliament.

Howitt, R. (1989) 'A different Kimberley: Aboriginal marginalisation and the Argyle Diamond Mine', *Geography* 74, 3: 232–8.

—— (1991) 'Aborigines and gold mining in Central Australia', in J. Connell and R. Howitt (eds), *Mining and Indigenous Peoples in Australasia*, Sydney: University of Sydney Press.

—— (1992a) 'Weipa: industrialisation and indigenous rights in a remote Australian mining area', *Geography* 77, 3: 223–35.

—— (1992b) *Aborigines, Mining and Regional Restructuring in Northeast Arnhem Land*, Working Paper 10, Economic and Regional Restructuring Research Unit, University of Sydney.

Howlett, D., Hide, R. and Young, E.A. (1975) *Chimbu: Issues in Development*, Canberra: Development Studies Centre, ANU.

Hunter, M., Hitchcock, R. and Wyckoff-Baird, B. (1990) 'Women and wildlife in Southern Africa', *Conservation Biology* 4, 4: 448–51.

Iglauer, E. (1966) *Inuit Journey*, Seattle: University of Washington Press.

Industry, Science and Technology Canada (ISTC) (1988) *Report on the Consultation Process on DRIE Native Economic Programs*, Ottawa: DRIE.

283

International Union for the Conservation of Nature and Natural Resources (IUCN) (1970) 'Resolution number one: definition of a park', in *Proceedings of the Tenth General Assembly*, New Delhi, India, 1969, Gland, Switzerland: IUCN Publications, New Series No. 27.

Ironside, R.G. (1988) 'Marginality and regional disparity in Canada', unpublished paper presented at IGU Congress, Sydney.

Isaacs, A. (1988) 'Different rules for different artists', *Inuit Arts Quarterly* 3, 1: 7–9.

ISTC (Canada) (1989a) *Sectoral Profiles: NEDP Element III, SARDA Commercial Undertaking and SARDA Primary Producing Activities*, Ottawa: ISTC.

ISTC (1989b) *Approved Projects by SIC Division: June 1, 1983–August 12, 1989: NEDP Element III*, Ottawa: ISTC.

Jull, P. (1991) *Australian Nationhood and Outback Indigenous Peoples*, Discussion Paper No. 1, Darwin: NARU, ANU.

—— (1992a) *The Constitutional Culture of Nationhood, Northern Territories and Indigenous Peoples*, Discussion Paper No. 6, Darwin: NARU, ANU.

—— (1992b) *An Aboriginal Northern Territory: Creating Canada's Nunavut*, Discussion Paper No. 9, Darwin: NARU, ANU.

Jull, P., Mulrennan, M., Sullivan, M., Crough, G. and Lea, D.A.M (1994) *Surviving Columbus: Indigenous Peoples, Political Reform and Environmental Management in Northern Australia*, Darwin: NARU, ANU.

Kann, U., Hitchcock, R. and Mbere, N. (1990) *Let Them Talk: a Review of the Accelerated Remote Area Development Program*, Report to MLGL and NORAD.

Keith, R.F. and Saunders, A. (eds) (1989) *A Question of Rights: Northern Wildlife Management and the Anti-harvest Movement*, Ottawa: Canadian Arctic Resources Committee.

Keith, R.F. and Wright, J.B. (eds) (1978) *Northern Transitions Volume II*, Second National Workshop on People, Resources and the Environment North of 60°, Ottawa: Canadian Arctic Resources Committee.

Kesteven, S. (1984) 'Alcohol and family life: the social impact of mining', in *Aborigines and Uranium*, Report on the Social Impact of Uranium Mining on the Aborigines of the Northern Territory, Canberra: AIAS, AGPS.

Kimberley Land Council (1991) *The Crocodile Hole Report*, Derby: Kimberley Land Council.

Knudtson, P. and Suzuki, D.T. (1992) *Wisdom of the Elders*, North Sydney: Allen & Unwin.

La Rusic, I., Bouchard, S., Penn, A., Brelsford, T., Deschenes, J.G. and Salisbury, R. (1979) *Negotiating a Way of Life: Initial Cree Experience with the Administrative Structure arising from the James Bay Agreement*, Montreal: DIAND.

Lea, D.A.M.and Curry, G. (1988) 'A Maprik journey: backwards or forwards in time? Cash cropping among the Abelam', in J. Hirst, J. Overton, B. Allen and Y. Byron (eds), *Small-Scale Agriculture*, Canberra: Commonwealth Geographical Bureau and Dept. of Human Geography, ANU.

Lea, J and Zehner, R. (1986) *Yellowcake and Crocodiles,* Sydney: Allen & Unwin.

Ledgar, R. (1986) 'Bow River Report', unpublished report to East Kimberley Impact Assessment Project.

Lee, R. (1972) '!Kung spatial organization: an ecological and historical perspective', *Human Ecology* 1, 2: 125–47.

—— (1979) *The !Kung San: Men, Women and Work in a Foraging Society*, Cambridge: Cambridge University Press.

Lennard, C. (1990) 'The Warlukurlangu artists experience', in J. Altman and L.Taylor (eds), *Marketing Aboriginal Art in the 1990s*, Canberra: Aboriginal Studies Press.

Loermans, H. (1992) 'Sustainable development in the Kalahari from a gender perspective', unpublished paper delivered at AAA conference, San Francisco, California, December 1992.

Long, J. (1970) *Aboriginal Settlements*, Canberra: ANU Press.

Lonner, T.D. (1986) 'Subsistence as an economic system in Alaska: theoretical observations and management implications', in S.J. Langdon (ed.), *Contemporary Alaskan Native Economies*, Latham MD: University Press of America.

Mabogunge, A. (1989) *The Development Process: a Spatial Perspective*, London: Hutchinson.

Macartney, W.J.A. (ed.) (1988) *Self-determination in the Commonwealth*, Aberdeen: Aberdeen University Press.

McGrath, A. (1987) *Born in the Cattle*, Sydney: Allen & Unwin.

Macpherson, J. (1978a) 'The Pine Point Mine', in E.B. Peterson and J.B. Wright (eds), *Northern Transitions*, Vol. 1, Ottawa: CARC.

———— (1978b) 'The Cyprus Anvil Mine', in E.B. Peterson and J.B. Wright (eds), *Northern Transitions*, Vol. 1, Ottawa: CARC.

Macquarie, L. (1816) Proclamation on the settlement of the colony of New South Wales.

Maddock, K. (1983) *Your Land Is Our Land: Aboriginal Land Rights*, Ringwood, Victoria: Penguin.

Manuel, G. and Posluns, M. (1974) *The Fourth World: an Indian Reality*, New York: Macmillan.

Marshall, L. (1976) *The !Kung of Nyae Nyae*, Cambridge, MA: Harvard University Press.

Meehan, B. (1982) *Shell Bed to Shell Midden*, Canberra: AIAS.

Merritt, J., Fenge, T., Ames, R. and Jull, P. (1989) *Nunavut: Political Choices and Manifest Destiny*, Ottawa: Canadian Arctic Resources Committee.

Merritt, J. and Fenge, T. (1990) 'The Nunavut land claims settlement: emerging issues in law and public administration', *Queens Law Journal* 15, 2: 255–77.

Middleton, R.M and Francis, S.H. (1976) *Yuendumu and Its Children – Life and Health on an Aboriginal Settlement*, Canberra: AGPS.

Miller, M. (chairman) (1985) *Report of the Committee of Review of Aboriginal Employment and Training Programs*, Canberra: AGPS.

Mogwe, A. (1992) *Who Was (T)HERE First?*, Report to Botswana Christian Council, Occasional Paper 10, Gabarone.

Momsen, J. (1991) *Women and Development in the Third World*, London: Routledge.

Momsen, J. and Kinnaird, V. (eds) (1993) *Different Places, Different Voices: Gender and Development in Africa, Asia and Latin America*, London: Routledge.

Momsen, J. and Townsend, J. (1987) *Geography of Gender in the Third World*, London: Hutchinson.

Morony, R. (1991) 'The Community Development Employment Projects (CDEP) Scheme', in J.C. Altman (ed.), *Aboriginal Employment Equity by the Year 2000*, Academy of Social Sciences in Australia/Center for Aboriginal Economic Policy Research, Canberra: ANU.

Morrison, W.R. (1983) *A Survey of the History and Claims of the Native Peoples of Northern Canada*, Ottawa: DIAND.

Mowbray, M. (1986) 'State control or self-regulation?: on the political economy of local government in remote Aboriginal townships', *Australian Aboriginal Studies* 2: 31–9.

Muller-Wille, L. (1978) 'Cost analysis of modern hunting among the Inuit of the Canadian central Arctic', *Polar Geography* April–June: 100–14.

Myers, M. (1988) 'Who will control, who will pay', *Inuit Art Quarterly* 3, 1: 3–11.

Myrdal, G. (1963) *Economic Theory and Underdeveloped Regions*, London: Methuen.

Nasogaluak, W. (1983) 'Reindeer herding and the reindeer industry in NWT', unpublished papers, Parts I–III, presented to Beaufort Environmental Review Panel.

Nasogaluak, W. and Billingsley, D. (1981) 'The reindeer industry in the Western Arctic of Canada: problems and potential', unpublished paper, Inuvik.

Nathan, P. and Liechleitner, D. (1983a) *Health Business*, Central Australian Aboriginal Congress (CAAC), Alice Springs: Kibble Books.

—— (1983b) *Settle Down Country*, CAAC, Alice Springs: Kibble Books.

Native Economic Development Program (1985) *Native Women and Economic Development*, Task Force Report, Ottawa: DRIE.

North Kimberley Land Conservation District (NKLCD) (1991) 'Feral donkey control in the North Kimberley', *NKLCD Newsletter*, Derby.

Nyae Nyae Development Foundation of Namibia (NNDFN) and Nyae Nyae Farmers Cooperative (NNFC), (1992) *Progress Report, 1991/92*.

O'Faircheallaigh, C. (1986a) 'A framework for northern economic development', paper delivered at NADC, Kununurra: North Australia Development Conference.

—— (1986b) 'The economic impact on Aboriginal communities of the Ranger Project: 1979–1985', *Australian Aboriginal Studies* 2: 2–14.

—— (1988) 'Uranium royalties and Aboriginal economic development', in D. Wade-Marshall and P. Loveday (eds), *Northern Australia: Progress and Prospects, Vol. 1 – Contemporary Issues in Development*, Darwin: NARU, ANU.

—— (1991) 'Resource exploitation and indigenous people: towards a general analytical framework', in P. Jull and S. Roberts (eds), *The Challenge of Northern Regions*, Darwin: NARU, ANU.

Paine, R. (1977) 'The path to welfare colonialism', in R. Paine (ed.), *The White Arctic: Anthropological Essays on Tutelage and Ethnicity*, Social and Economic Papers No. 7, Memorial University of Newfoundland, Newfoundland: University of Toronto Press.

Palmer, I. (1988) *Buying Back the Land; Organisational Struggle and the Aboriginal Land Fund Commission*, Canberra: Aboriginal Studies Press.

Park, R.E. (1937) 'Introduction', in E.V. Stonequist, *The Marginal Man*, New York: Charles Scribner's Sons.

Perkins, J. and Thomas, D. (1993) 'Environmental responses and sensitivity to permanent cattle ranching, semi-arid Western Central Botswana', in D.S.G. Thomas and R.J. Allison (eds), *Landscape Sensitivity*, Chichester: John Wiley & Sons.

Peterson, E.B. and Wright, J.B. (1978) *Northern Transitions*, Vol. I, Northern Resource and Land Use Policy Study, Ottawa: Canadian Arctic Resources Committee.

Polar Gas, (1985) *Polar Gas Training Program: a Discussion Paper*, Calgary/Toronto: Polar Gas.

Ponting, J.R. (ed.) (1986) *Arduous Journey: Canadian Indians and Decolonization*, Toronto: McClelland & Stewart.

Porter, D., Allen, B. and Thompson, G. (1991) *Development in Practice: Paved with Good Intentions*, London and New York: Routledge.

Press, A.J. (1987) 'Fire management in Kakadu national park: the ecological basis for the active use of fire', *Search* 18, 5: 244–8.

Pretes, M. (1988a) 'Underdevelopment in two norths: the Brazilian Amazon and the

—— (1989) 'Approaches to planning in native Canadian communities: a review and commentary on settlement problems and the effectiveness of planning practice', *Plan Canada* 25, 1: 56–67.

—— (1993a) *The ATSIC Aboriginal Community Development Planning Program in Northern Australia: Approaches and Agendas*, Discussion Paper No. 16, Darwin: NARU, ANU.

—— (1993b) *Regional Planning by ATSIC Councils: Purpose, Process, Product and Problems*, Discussion Paper No. 18, Darwin: NARU, ANU.

Wonders, W.C. (1987) 'The changing role and significance of native peoples in Canada's Northwest Territories', *Polar Record* 23, 147: 661–71.

Wonders, W.C. (ed.) (1988) *Knowing the North: Reflections on Tradition, Technology and Science*, Occasional Publication No. 21, Boreal Institute for Northern Studies, Edmonton: University of Alberta.

World Commission on Environment and Development (WCED) (1987) *Our Common Future*, Oxford: Oxford University Press.

Wuttunee, W.A. (1988) 'Competing goals and policies of Alaska native regional corporations', unpublished MA thesis, University of Calgary.

Young, D.E. (ed.) (1988) *Health Care Issues in the Canadian North*, Occasional Publication No. 26, Boreal Institute for Northern Studies, Edmonton: University of Alberta.

Young, E.A. (1982) 'Aboriginal community stores: a service "for the people" or "by the people"', in P. Loveday (ed.), *Service Delivery to Remote Communities*, Darwin: NARU, ANU.

—— (1984) *Outback Stores: Retail Services in North Australian Aboriginal Communities*, Darwin: NARU, ANU.

—— (1987a) 'Resettlement and caring for country: the Anmatyerre experience', *Aboriginal History* 11, 2: 156–70.

—— (1987b) 'Commerce in the bush: Aboriginal and Inuit experiences in the commercial world', *Australian Aboriginal Studies* 2: 46–53.

—— (1988a) *Aboriginal Cattle Stations in the East Kimberleys: Communities or Enterprises*, EKIAP Working Paper No. 21, Canberra: CRES, ANU.

—— (1988b) 'Land use and resources: a black and white dichotomy', in R.L. Heathcote and J.A. Mabbutt (eds), *Land, Water and People: Geographical Essays in Australian Resource Management*, Sydney: Allen & Unwin.

—— (1988c) 'Striving for equity; Aboriginal socio-economic transformations and development in the 1980s', *Geoforum* 19, 3: 295–306.

—— (1988d) 'Aborigines and land in Northern Australian development', *Australian Geographer* 19, 1: 105–16.

—— (1988e) 'Aboriginal economic enterprises: problems and prospects?, *Northern Australia: Progress and Prospects, Vol. 1 – Contemporary Issues in Development*, Darwin: NARU, ANU.

—— (1991a) 'Comparative issues in remote area aboriginal development: Australia and North America', in I. Moffatt and A. Webb (eds), *North Australian Research: Some Past Themes and New Directions*, Darwin: NARU, ANU.

—— (1991b) 'Marginal land and marginalised people: how indigenous people can contribute to economic development in Australia and Canada', *Journal of Australian-Canadian Studies* 8, 2: 75–98.

—— (1991c) 'Australia, Canada and Alaska: land rights and Aboriginal enterprise development', in P. Jull and S. Roberts (eds), *The Challenge of Northern Regions*, Darwin: NARU, ANU.

—— (1992a) 'Aboriginal land rights in Australia: expectations, achievements and implications', *Applied Geography* 12, 2: 146–62.

—— (1992b) 'Hunter-gatherer concepts of land and its ownership in remote Australia and North America', in K. Anderson and F. Gale (eds), *Inventing Places: Studies in Cultural Geography*, Melbourne: Longman Cheshire.

—— (1993) *Support Organisations for Aboriginal Community Stores: the Arnhem Land Progress Association and Its Counterparts in Central Australia and the North American Arctic*, Darwin: NARU, ANU.

Young, E.A., Crough, G. and Christopherson, C. (1993) *An Evaluation of Store Enterprises in Aboriginal Communities*, Darwin: NARU, ANU.

Young, E.A. and Doohan, K. (1989) *Mobility for Survival: a Process Analysis of Aboriginal Population Movement in Central Australia*, Darwin: NARU, ANU.

Young, E.A., Ross, H., Johnston, J. and Kesteven, J. (1991) *Caring for Country: Aborigines and Land Management*, Canberra: ANPWS/ATSIC.

Young, E.A. and Ross, H. (1993) 'Sustainable development planning for Aboriginal Communities', *Australian Aboriginal Studies*, 1: 91–2.

Yu, P. (1994) 'The future of the Kimberley', in P. Jull, M. Sullivan, M. Mulrennan, G. Crough and D. Lea (eds), *Surviving Columbus: Indigenous Peoples, Political Reform and Environmental Management in N. Australia*, Darwin: NARU, ANU.

Yukon Government (1988) *The Caribou Are Our Life*, Whitehorse: Yukon Government.

Zarsky, L. (1990) *Sustainable Development: Challenges for Australia, Commission for the Future*, Canberra: AGPS.

OFFICIAL GOVERNMENT FILES

Department of Aboriginal Affairs (DAA) (1977/166) Ti Tree Station.

PERSONAL COMMUNICATIONS

Art Dedam, Economic Development Officer, Assembly of First Nations, Ottawa, Canada.

Robert Higgins, Project Officer, Inuit Tapirisat of Canada, Ottawa, Canada.

Stephen Horn, Financial Manager, Kahnawake Council, Montreal, Canada.

Paul Kariya, Department of Indian and Northern Development, Ottawa, Canada.

INDEX